5/16/11 For ... confirmation
of his
Thanks for
been a pri...
and to share ...
Neil and Joan

PERFECT MISTAKES

Neil H. Mendelson

2005

ISBN-13: 978-0-9765345-1-8
ISBN-10: 0-9765345-1-7

90000

9 780976 534518

This book is dedicated to those who accepted me for what I was with no malice for what I was not. Its purpose is to inform those who knew me what I was really like beneath the façade they came to believe was the person. The story isn't meant to shock or even surprise but rather to set the record straight. The factors that shaped my existence probably are not uncommon although the way they played out in my particular case may be. Sure it was a bumpy road but the lumps I took never diminished my resolve to find the better place, the better life. The great joy now is seeing the progressive achievement of those goals by my children and grandchildren. The corner may indeed have been turned.

It would not be honest to leave unsaid the fact that had not my wife Joan rescued me from the emptiness, there would have been nothing to write about here.

1. The man who was four parts.

This is a true story about a professor emeritus, what that means, and how he got that way. If you have never seen a professor emeritus before this reading, relax you're in the majority. And don't worry about meeting another one either; most of them are nearly invisible. They can't or do not wish to profess anymore. Emeritus is supposed to mean, "retired from active service but retaining one's title". When "emeritus" follows the word "professor" however its true meaning is actually closer to "a basket case". It's not that simple though. These basket cases come in many varieties. More flavors than ice cream in fact. The one I know, and about whom you will hear is not the plain vanilla form. He is a person, who from the outside appears to be totally in command of his own fate, as well thought out as a fine Swiss watch, and as accomplished as anyone could hope to be. In truth he is not any of these. What he has achieved is nothing short of a miracle. He is way outside of the mainstream of every category. How could anyone have lived such a life? The objective of the story then is to provide a rational explanation for both his differences and the way he accommodated them.

All the names you see in this book are not fictitious. They portray the real people who did or wrote the things presented. The settings in which events unfolded are also the real ones, not just stage sets providing the space upon which the characters prance about. So too are the times at which events are described. Everything does not always go in strict chronological order. Keep that in mind. The reason for the discontinuities is that particular anecdotes frequently entail events that happened years apart but are linked in terms of their significance. The time gaps between some of these events reflect the long overall time span covered by the story and the lifestyle of scholars who work in an international sphere. Relationships, scientific and social, stretch out in time and distance but their perceived stopping and starting points are seamless in continuity. How the mind can do this is anybody's guess but I suspect it has something to do with the fact that people like our professor emeritus are rather isolated in the world. When they find others in their small circle they hang on to them rather tightly, to hell with space and time. There is another side to it as well. The profession itself has its own circle and those inside it, like them or not, will reappear again and again just as weeds in your garden. If you don't deal with them each time you can, you will pay the price later.

Our central character, Neil H. Mendelson, (NHM) made his livelihood and his reputation as a scientist/professor, then secondarily as a musician, but he also did a lot of other things as well. He was a Captain in the Army, he painted in oils and watercolors, wrote poems to mark life's events, was a lavender farmer for a while, and frequently rattled the tranquility of relatives, friends, and strangers too. The spectrum of things he did really didn't fit together in a logical way. Nor did the unusual way in which he met challenges and competition. Whatever it was he did, his way of doing always differed from the way it was usually done. And most of the time he ended up with a unique product as surprising to him as it was to everybody else. He clearly followed a different path through life, one that did not pass through most of the places others have been. Later when he himself recognized this it became clear to him that for every wonderful thing accomplished throughout his journey there was a price that had to be paid. And in the end he realized just how much he had missed in life. That is when it became important for him to try to figure out what the reasons were for his always taking the path of most resistance rather than least resistance. It was as if he always had to swim against the current. Going up river not only saps your energy but it inevitably leads to a dead-end. Even fish know this. NHM should have known from his exhaustion right from the start that he was pointed in the wrong direction. And from his frustrations how swift the current was. And from his isolation it should have been clear to him how few there were who had his orientation. All the big fish weren't going upstream with him. Where were they going he wondered? Eventually the questions became how he got into this stream in the first place, what if he got trapped in a pool, would he ever be able to get out and if not what would he do there?

The story will be told by one of the four internal characters that make up professor emeritus NHM, the character referred to as the "Observer". The Observer had the responsibility for noting, organizing and storing all information. Sensory inputs of every nature went to him, including those generated from within by the other three characters.

Whatever went into the Observer's repository was always freely available to be examined or used as necessary, but there was a strict rule about the information; the original could never be taken out or changed. This assured accuracy is what makes the Observer the ideal one to relate the tale. The second of the four characters, the "Creator", was responsible for the unique nature of the person. To him goes the credit for figuring out how to survive, what to do, and for interpreting what things meant. The Creator spoke often to the Observer taking from him and giving to him a flow of things that would have made a terrible whoosh had they been in the form of a moving fluid rather than thought. When you confront the unexpected or totally audacious ideas in this recounting of a life their origin must have been the Creator. The doing of things fell under the control of yet another character, the "Performer". Translating thoughts to actions were his responsibility and he worked with elan. He loved his job as will be evident. Had there not been some constraints he may very well have blown most everything out of proportion resulting in chaos beyond redemption, chaos that could have sunk the whole ship. The Performer made life rather enjoyable for his three partners. If alone as a pair, the Performer and the Creator could easily have run amok, as the Observer well knew. The credit for keeping them from ruining everything goes without question to a woman named Joan, NHM's wife, and to the forth-internal character known as, the "Guardian" of the core. What the Guardian protected was assembled from life's experiences and the works of his three partners. It was a set of beliefs that formed the orientation for everything in life without which there could be no understanding, or hope, and no future. The Guardian had power to curtail the Performer, and to make sure the Creator paid adequate attention to the core before he rushed so far beyond it that he jeopardized the whole. One thing must be kept in mind about the Guardian; he bore no responsibility for anything in the core that might have been wrong. His job was to protect the edifice as it stood. It was of no concern to him that the whole thing might have been constructed from *"perfect mistakes"*.

Observer, Creator, Performer, and Guardian had a lot of material to work with but not everything. The totality of information available to them was rather narrowly focused. Observer wished for more and for some editing mechanism beyond Creator's interpretation of everything, some means to verify facts and interpretations before they became part of the core. But he had no help in this. Instead Creator pushed ahead and wanted no distractions. He had good reasons for doing so. Survival was in his hands, he knew it and could barely manage meeting its demands. There just wasn't enough time to check for accuracy. Creator was pragmatic when he had to be. But most of the time his proclamations were definitive and he was confident that he would get it right, if not immediately then eventually. Part of his confidence came from the great trust he had in what Performer could accomplish. Creator failed to appreciate however that there is a class of mistakes in this world that can never be corrected. These are the *"perfect mistakes"*. Just as time lost can never be regained so long as time's arrow always points to the future, perfect mistakes are barred from correction by some kind of force not really understood but just as real. The unfortunate thing for NHM was that Creator's distraction with survival came at precisely the time in NHM's life when his core had to be assembled. Without knowing it then ten perfect mistakes were made and they became foundations of his core being. Guardian then took over and life's options were foreclosed. Creator had inadvertently made his and Performer's jobs infinitely more difficult. Their struggles and achievements are much of what follows. When coupled with the details of the environments in which events took place, the role chance played, and what it was like during the period that the core became established you will see that Creator actually deserves much praise. Ten perfect mistakes could easily have been a thousand.

Just as I, Observer, archived everything, Performer did likewise, keeping as much paper as he could for as long as he could. So, where I, Observer, can tell you something from my recollections Performer can show you from his records precisely what had been written about it and by whom. Recollections and records together will establish the facts. Then I will tell you how Creator has interpreted them. The four of us on the inside believe that a rational explanation can be made for the form and substance of NHM's existence from the evidence presented. Though different, what NHM did is not unfathomable. Quite the contrary; it is at least to us the most logical outcome possible given his circumstances. In later life Creator spent considerable time pondering what role genes might have played, whether they might be held responsible thereby exonerating any of us from pushing NHM so far out of the mainstream. Could this be nothing more than genetic information gone haywire that NHM inherited and that he may pass to his offspring? Think of the fish that go against the current, reach the dead end upriver and reproduce there just before they die. Their offspring ride the current out with no clue that someday they too will have to fight the current. It is in their genes. They have no other choice. Could this be it for NHM as well? Were Creator's perfect mistakes predetermined in NHM's genes? That would certainly take the heat off Creator wouldn't it? It sounds reasonable to me the Observer, but Creator himself isn't willing to pass the buck. He's a tough character willing to accept responsibility even when it isn't yet certain that he deserves to be blamed. Either way we all agree that ten perfect mistakes are the heart of the matter. Or do we?

When Creator finally became aware of the consequences of his doings he asked me, Observer, for a summary of the results, something with enough detail so he could, "connect the dots" as he put it, which I took to mean that he also wanted things from Performer's archive. This is what the two of us provided. Our host, NHM, and his colleagues in science were perplexed by the fact that his work had been so totally different from theirs. Not just one or two things but the entire body of his productivity, a volume that fills nine hundred plus pages of a book entitled, "Collected Papers of Neil H. Mendelson" a copy of which Performer can provide. Likewise, his colleagues in music who considered his life as a scientist/professor his "day job" wondered how he managed the heavy performance schedule demanded by the symphony orchestra he played in as well as other concerts he was hired to perform. Note as well that the instrument he plays, the double bass, is not exactly the most common instrument. Classical bass players are themselves an unusual minority. Then out of nowhere, he became a lavender farmer. It doesn't make sense. Be it in the laboratory, lecture hall, on the farm, in the concert hall, or even just in public, it did not matter where, our host, NHM, repeatedly did shocking things. He himself recognized something must have been wrong. Others agreed and add to their amazement the fact that NHM did not do or even like most of the things his generation did. Other people's actions, even their objectives seemed ludicrous to him. He was amused by the ridiculous ideas people had. Consequently he presented little challenges to them just to observe how they responded to something out of the ordinary. Take a look at this example provided by Performer. It is a question he wrote for one of the annual PhD qualifying examinations in genetics at the University of Arizona.

Doggerel named answer-me (DNA for short)

There was a naughty little gene
that lived inside a human being.
It traveled here it traveled there,
wherever it went it caused despair.

Why was this naughty gene so mean?
Why did it hurt the human being?
How could it go on doing that,
without itself going splat?

One day a man named know-it-all
sat down to think and to toil.
He asked himself, what can I do,
to drive this naughty gene from you?

Inside the wizard's mighty brain,
a clever thought turned into pain.
Ask the students what they think.
Give them a chance to swim or sink.

And so today I'm asking you.
What on earth can we do?
Is there a way to lick this thing,
or, must we go on suffering?

Not exactly something they could find in the back of a book or in his lecture notes. Fortunately no one became ill trying to answer it. He, on the other hand, enjoyed watching the "marbles in their minds" rattling around while they were trying to scratch out an answer. Could they have learned something from the question, he wondered? NHM always believed that an exam is one of the best ways to teach. Test anxiety has its benefits, he says. It gets the marbles moving. Taking it to the extreme, he then proposed that perhaps older people should be made to take a test every few months to get their Social Security checks. It might provide the gray power needed to save this country. There is too much emphasis on physical health these days he claims rather than intellectual health. And to rectify the matter this is what he has done. He has gone out into the world at large to help the ordinary people by administering not exactly tests but what he calls, "mental challenges" to them. This he believes jogs their minds. Of course he cannot predict the outcome but nine times out of ten he learns something from it and he enjoys the way they respond very much. The subjects are not uniformly appreciative of his help. Here are some examples. At some level Creator must be responsible for formulating these unusual behaviors.

Take the case known as the black spider ploy. On a beautiful and peaceful day in his laboratory when everything was working, professor NHM got a telephone call from his wife, Joan. I'm in Costco she told him, looking at the Oriental rugs that are here for only this week. There are several that might be suitable for our farmhouse. Can you come over to look at three of them that I have picked out? A quick calculation told him that it would cost about 1 hour for him to do so. Do you think I can get back in an hour he asked her? Yes, probably she replied, so off he went. Of the three she had chosen they both agreed one was not as good as the other two so back it went onto the pile. Either of the other two would have been suitable. The question was which one to take. Back and forth they went making all the comparisons. The salesman, a young Arab man, tried to make some small talk pointing out things about each rug. NHM knew which one he preferred but he wasn't sure which one Joan favored. Finally out of the blue he said to the seller, "Do you know there is an ancient myth that says when you can't make up your mind which of two carpets to choose, a spider appears on the one you should avoid?" No, the seller replied, I haven't been in this country that long and I never heard about that. Then when NHM was certain which of the two his wife really wanted, he got down on the other rug and began examining it very closely. The salesman got down too. After a few moments the two came head to head. The salesman looked him in the eyes and said I don't see the spider, as if to mock the myth. NHM glanced back at him looking very stern but said nothing as he pointed down. There on the carpet was the black spider, neatly planted from his pocket. When the salesman saw it, he nearly fainted. In his confusion NHM quickly retrieved the spider, and said to Joan, "the black spider has appeared here so it must be the other rug we buy." And they did. The next day NHM's daughter returned to Costco to see if she could find some suitable rugs for her sons' rooms. She told the salesman that her parents had purchased a rug from him the day before and she wanted to see the other two that they didn't buy. He looked at her for a moment then said, "The man with the hat and the spider". "That would have been him" she replied without giving it a second thought although she had no idea what the spider had to do with it. Even his grown children can recognize his perfect mistakes phenotype. But where did the spider come from? To this Creator then added, yes I remember the incident well. NHM found the plastic spider in his pocket on the way into Costco. Without realizing it he had carried it around since Halloween. Once Creator knew it was there it was just a matter of time until he found the opportunity to use it. Then Performer took over. There you have it.

What about the thumb incident in Paris Observer asked Creator, do you remember it? Not fully, tell me about it please, he responded. It was when professor was working at the Pasteur Institute on one of his most important discoveries, bacterial macrofibers. The physician working in the laboratory next to his let out a frightening scream and ran from his laboratory one afternoon. NHM rushed into the hall to see what happened but was too late, the poor man had run out of the building screaming. The French scientists who witnessed it seemed very calm as if they knew what it was all about. Indeed they did. They had planted in his laboratory a horrible rubber thumb that was made to look as if it had been ripped from someone's hand in a terrible accident. He, they knew, was made ill by it. His departure was part of the normal drill after he encountered it, providing time for the perpetrators to retrieve it for use another day perhaps on someone else. The following evening the thumb made its appearance in professor NHM's bath towel at his apartment and scared the life out of him. How did it get there? Well, they taped it onto the inside of the lid to his lunch box. He never noticed it there, even though others on the Metro kept staring at his box, so when he got home he simply dropped off the box in the kitchen where Joan would wash it and refill it for the next day. When she recovered from the shock of finding it, she thought NHM was responsible so retaliated by hiding it in his towel. The perpetrator's dual success required a robust response on NHM's part. What he did, did the job. Yes I am starting to remember it now Creator interjected. Wasn't it used in his lecture as a pointer? Yes it was. Professor knew that exposing their little horror in public would be something that would never be done given the ethics of the institution. No one would be able to laugh in public at a thing like that so what would they do he wondered. The stick to which it was attached was fabricated for him in the Pasteur shop presumably for another purpose. Creator planned how to get it into the lecture hall all wrapped up in paper sheets thought to contain his research data, and just how to use it as a pointer. When the thumb first appeared in the darkness to point out details on the projected graph it looked as if his own thumb had grown five feet long from his body. Those who knew what the thumb was, gasped. The others sat bewildered trying to focus on the science and the thumb at the same time. Owners of the thumb fought laughter with all their energies. Professor knew they could not reveal their identities, so he stretched out the lecture as long as he could to settle his debt. Afterwards back in the department pandemonium broke loose. All their pent up hysteria poured out. They brought the man from the shop who fabricated the stick to show him its true purpose. Professor returned the thumb to its owners. His lab neighbor, the physician, congratulated him on having gotten back at them. Another senior French professor was angry with NHM for the rest of the year. Yes Creator said we jogged a lot of minds that afternoon. To which observer added, indeed and in the very setting where NHM discovered helix hand inversion in bacterial macrofibers, a truly unique process in the microbial world. The very kind of handedness that Pasteur had found in crystal structure was now known to exist in bacterial structure. More will be said about that later and how awkward the finding was. The handedness that Pasteur and NHM had worked on had nothing to do however with the hand from which the thumb had been modeled!

I must remind you as well, Observer told Creator, of a similar incident involving a pointer that took place when NHM was in graduate school. He was in his early twenties then in 1961. I shall not give all the details here but will return to them in another context. Recall it was when he had to present a lecture as part of his Ph.D. requirement in genetics, just before the start of his research project. In his talk he used a bull penis as a pointer. Never mind for now how he obtained such a thing. To use it as he did was certainly in the worst possible taste. He knew that, yet he went right ahead with his outlandish prank. What might have motivated such behavior? I am going to try to figure that out, Creator answered. From what I have already seen it appears that our man somehow got the wrong basic set of ideas about how things work and probably also about his place in life.

You realize, I hope, that professor received his fair share of warnings about what not to do in formal settings that might embarrass his hosts. Later in his career he had been granted privileges at two Cambridge Colleges on the basis of his scientific achievements and proper introductions. There when he was turned loose at high table or in the combination room it was a challenge for him not to mock the shocking degree of pomposity that lives among the academically privileged of Great Britain. What is the purpose of such institutions he wondered after witnessing the degree of backbiting that went on in them. A very accomplished and highly rewarded colleague of his once revealed to him that he viewed the major benefit of affiliation to be the parking space it provided him in the center of things. NHM said he would have guessed it to be the food they served. It was on a par with the finest of London restaurants. The man responsible for the food service, a fellow in the college, was also a famous legal scholar. Where he acquired his outstanding culinary skills is anybody's guess. When NHM dined there as a guest he was seated next to the legal scholar. Fortunately their discussion did not focus on food. Years later NHM learned that he was very sensitive to anyone mistaking that his talents as the steward were on a par with his accomplishments as the scholar. Pity those who did not know that. If you don't know the local culture best to keep your mouth shut Performer concluded, but Creator often pushed NHM beyond this checkpoint.

It seems to me given what we already know, Creator continued, that NHM came to rely largely if not entirely upon his own innate abilities. He valued leadership more than followership. And he may have jumped to the conclusion that there are just rewards for effort and achievement. Could he have believed that science is a purely objective pursuit? Yes he did Observer told him and went on to say that NHM thought that what he had been taught was true. And, he strongly believed that it was possible to work by yourself independently and not have to rely on others for your own accomplishments. Most importantly, he assumed that those responsible for him would act responsibly. These mistaken beliefs matured into the ten perfect mistakes that became firmly imprinted with the help of the Guardian. They set the boundaries for NHM's' life plan. To them he owes his primary career as a research scientist and professor, his secondary career as a musician, the fact that he never did anything of practical value, and the fact that nothing has ever come easily. We still need to learn though why he knew so little about power and people and why that meant barriers at every turn in life. As NHM himself thought, "when you do not do what everyone else does, what do you do? What can you do?" A good option might have been to enjoy life, but first you must survive.

I spent a lot of energy on survival, Creator said, so I am anxious to hear Observer's version of the details. Here are some. NHM survived the disruption of the Second World War, life on the streets in New York City, being uneducated by the city's public education system, a dysfunctional family, and an Ivy League college education during the period when seven out of ten students admitted failed to graduate! He was as out of place going to graduate school in the mid-west as a duck landing in the desert. Time there appeared to have been going backwards. It must have been a trick of relativity however. Eventually he finished his Ph.D. and moved on never to return to the "heartland of America". His next challenge was active duty in the U.S. Army. It was not an ideal time for an officer to serve given the situation in Vietnam. He had little choice however when he was ordered by the Department of the Army to report for duty to his branch school where he would be educated in preparation for a duty assignment. In his branch class there were two officers with Ph.D.'s, NHM being one of them. They were first lieutenants. The other students were second lieutenants. Higher rank meant nothing there. All student officers were grouped together to learn the skills of war and become physically fit to carry them out. Midway through his branch course the President of the United States was assassinated in Dallas. During war games scenarios before that the student officers had been fooled into thinking that real events had taken place that were in fact only fabrications of the learning plan, hence the initial notice that the President was dead was met with much skepticism. The leadership of the school attempted to convince them that the announcement was real by bringing small television sets into the buildings tuned to local news channels. The student officers were also allowed to leave the installation if they wished to so that they could confirm the truth elsewhere. A heightened state of alert was issued in view of the uncertainty as to whether a hostile foreign power might be involved. When it became clear that this was unlikely the instruction resumed leading up to graduation day and the fateful news that went with it: where the duty assignments would be for each officer. Assignment came from a special office in the Pentagon responsible for personnel operations. Their job was to find officers with individual skills that met the needs of particular military positions. NHM was fortunate not to receive a combat assignment. Little did he realize then that the job he had been given could have been as dangerous as going to Vietnam.

Toward the end of NHM's two-year tour of active duty, officers serving in his branch of the Army were being killed in the war at a rate that could not be met by the assignment of newly graduated officers from the branch school. There were two ramifications of this. The first was that the time in service requirements for promotion were reduced. NHM's promotion to captain came about six months before he normally would have been promoted. Second, officers were quickly being reassigned from non-combat jobs in the United States to Vietnam. An accident of timing prevented him from joining the others who were shipped out on short notice: NHM had too few days remaining on active duty for the transfer to be worthwhile from the Pentagon's perspective. He was permitted to finish out his service without having to do battle in Vietnam. Once he knew that his efforts were directed to making sure he didn't accidentally kill himself and in planning what to do if he did manage to live through it. How can I get back into academic life and research in pure science he asked himself? I do recall that, Creator said and that I was so exhausted from survival issues at the time that I had no really valuable suggestions to offer him. Performer simply solicited help from his former graduate school professors. One response he received set the course for his entire career. "Have you considered doing a postdoc in Europe?", wrote a famous old professor, Ralph Cleland. Indeed he hadn't but the idea sounded perfect. A round of additional letter writing eventually led to an outstanding offer in London. It came from the leading expert in bacterial genetics at that time, William Hayes, and had only one requirement: NHM would have to find support for his salary from the United States as all other Americans working in Hayes' research group had done. Now the short time left on active duty looked forward to for so long became a liability not a blessing. Would there be enough time to secure the necessary money? Performer went into high gear. He immediately drafted grant applications and submitted them to NIH and NSF. By then, 1965, Joan had given birth to twin daughters, it was a family of four that rode upon NHM's shoulders, not just his own survival. If he didn't get the money his opportunity, and possibly his research career, would be gone. Once discharged from active duty his income would also be gone. When notice that he received a postdoctoral fellowship award from a new program at the National Science Foundation arrived there was no time for celebration. All the details had to be worked out in the few remaining days of active duty. Performer worked himself to the bone, so to speak. Liberation day arrived and Captain Mendelson walked out the door after his debriefing knowing that his warrior days were behind him. Creator then began to focus on the beginnings of NHM's career as a scientist.

Thirty-five years later NHM met an older man who claimed he was one of his scientific grandfathers He said he was so pleased at how well things had worked out for NHM. This man, Dr. Bowen Dees was a mystery to all of us. Neither Observer nor any of his partners knew who he was or how he knew so many things about NHM's career. Dr. Dees himself provided the answer. He had been a high-ranking official at the National Science Foundation who was responsible for creating the NSF postdoctoral fellowship program. He knew NHM as one of the first recipients of his fellowships. He remembered that a Nobel Prize winning member of NHM's graduate committee strongly supported him, but wondered how NHM managed to get such a wonderful offer from London so quickly at the end of his military career. Performer told him that at the time we had no idea but after working for a while in the Microbial Genetics Research Unit (MGRU), as the group in London was known, we discovered the reason. Dr. William Hayes had himself been a medical officer in the British Army and later in the Indian Army when he decided to remain in India to complete some medical projects that he had initiated while serving there. NHM must have been the only scientist who had ever approached Hayes as a military officer seeking to resume a career in science, as Hayes himself had done after returning to Great Britain from India. Bill Hayes must have known how difficult the transition would be. Notified in 1993 that Bill Hayes was near death, Creator instructed Performer to write the following letter to him and his wife Nora, hoping it would get to them in Australia in time for him to see it. Luckily it did.

December 21, 1993

Professor and Mrs. William Hayes
527140 Pennant Hills Road
Normanhurst,
NSW 2076
AUSTRALIA

Dear Dr. and Mrs. Hayes,

I was sad to hear of Bill's ill health from Simon Silver who visited Tucson earlier this month. Although it has been more than 25 years since the days Joan and I and our children lived in London and I worked in Bill's MGRU at Hammersmith Hospital, we often think of those days and how wonderful it was. At the lab Bill created the atmosphere where we all really did think we would solve everything within a year or two! London was an exciting place, and your hospitality shall never be forgotten.

Joan and I wish we could have gotten to Australia at least once to see you both but unfortunately it wasn't possible. After a move to Tucson made necessary by my daughter's asthma we settled into lives constrained first by the children's needs then by our professions. Joan became a lawyer working in child protectives services after having been a geneticist to begin with. My work drew me into mechanics, engineering and mathematics more than bacterial genetics per se, although I still use *Bacillus subtilis* as the experimental system. At the same I did manage to become a symphony musician and Joan is trying to do the same although she's gotten a much later start.

We frequently pass through London on our way to Cambridge where I have been collaborating with a man in engineering for the past 10 years. In London we always go to our old neighborhood, and have on occasion gone by Hammersmith Hospital just to get a glimpse of past memories. Our year there in 1966 set the entire direction of my career. In a sense, Bill made everything possible for us. For that I am truly grateful.

Our best wishes to you both and regards for the new year.

Yours truly,

Neil H. Mendelson, PhD
Professor

Bill's widow Nora informed NHM that Bill loved it and that they had put it into their family archive for their grandchildren. Professor Simon Silver, a colleague who had also been a postdoctoral fellow in the MGRU, was in close contact with the Hayes'. He informed NHM in the first place that Bill was terminally ill and of Bill's death when it occurred. Simon knew also that our letter had gotten through in time and that parts of it were to be read at Bill's funeral. A copy of his email is shown below. Oh yes, the things said in NHM's letter to the Hayes' are indeed true. In retrospect all four parts of NHM were very pleased at how much they enjoyed what we had to say. It was a very small repayment for the kindness Dr. Hayes extended to NHM in 1966.

From: Simon Silver (1/10/94)
To: Neil Mendelson

Bill Hayes

Dear Neil,
Please ring me when you come in this morning. Office 312-996-9608, or if no one answers (the phones are funny) Lab: 312-996-3363, 15 feet away.

I rang Nora Hayes last night (Chicago time). She said that a wonderful letter from you had recently arrived and that she had read it to Bill, who seemed to understand - or at least understood the happiness of it. Then over the weekend Bill died. Your letter— or maybe parts of it—will be read aloud by Michael at his father's funeral Tuesday in Sydney.

I am sorry to convey such bad news, but maybe glad that we talked about Bill just in time for you to add confort for Nora. The good will and good feelings of Bill's students and postdocs are very important to her.
Simon

Simon told us that he later contacted the American Society for Microbiology asking for them to publish Bill's obituary but they refused to. Although world famous and a Fellow of the Royal Society, Bill was not a member of the ASM and thus not deemed worthy of the honor. As an alternate Simon suggested that we write a small paper about Bill and his accomplishments. He and NHM along with three other former MGRUers drafted the memoir. ASM News published it. Performer reminds me that a copy is included in NHM's Collected Papers book.

Telling you about Bill's death before you hear about the role he played in NHM's life puts the cart before the horse. Here is a capsule summary of those events. Alive and well, and a veteran to boot, NHM went off to London with his wife and two children in a state of euphoria. They climbed aboard the SS. United States with the anticipation of children headed to the best party ever. The trans-Atlantic sailing in October was as placid as a pleasure boat tour of New York harbor. Never mind that they learned upon arrival at Waterloo Station that the apartment they had rented before leaving had not been held for them and they had no place to go. Never mind that the Left-luggage office they were directed to, where they had to leave in storage all the things they brought with them until they could find a place to live, wasn't on the left side of the station. Where was it? Left meaning left behind, not a location. How could they have known that? It didn't matter. This was the party they dreamed of. Of course it never occurred to NHM that if he didn't make a significant discovery during his stay in England he might not be able to get back to a job in the United States. He had no mentors looking after his career in science who could help him get a job. The two-year period that he spent in the military was not a particularly attractive component of his curriculum vitae, especially to those in the academic community. To most academicians military service was not viewed as "the right thing to have done". So his future was staked on making a significant discovery. When he did eventually find something of scientific merit he assumed everything would take care of itself. It didn't. He had no idea of how to find a job as an assistant professor or any other kind of scientific position for that matter. Even Creator was out of his element in this. A chance transfer of mail that he held for a former professor of his who was passing through London on the way elsewhere solved the problem. It came about because the visiting guest told him that a mutual colleague of theirs was about to move to a new university campus and was looking for faculty to staff a department that he was going to build and chair.

In 1967 NHM became a newly minted assistant professor in this department. He pursued science for the rest of his life in the isolated world of the laboratory and the lecture hall. The grueling fight to sustain that lifestyle was not without its rewards and its humor. Nor was it without a price that his wife and children had to pay, something it took him a long time to recognize. In the telling that is to follow I will give you some idea of what it has been like working in two state universities and in the competitive world of scientific research at an international level. NHM managed as well to find his way into a symphony orchestra twenty five years ago. Performer had a new life of his own there. Creator recognized a similarity between getting up before people to perform a concert and getting up to give a scientific lecture but never appreciated the striking parallel faced by both scientists and musicians until NHM's experiences in both provided him with the information. In both professions the odds are strongly against success. For those who must earn a living to survive these are not the best choices one can make for a career. For those who don't have to worry about such things, these may be the best choices one could make for a life. Science and music share beauty that is a rare commodity in this world.

The Guardian of the core will not be shocked by what follows here but others may well be. The beauty of science and music has its costs. The institutions in which they are practiced are not perfect nor is the culture associated with them. Parents reading this who are about to send their children off to fend for themselves for the first time in institutions like the ones NHM has experienced from the inside may not be calmed by the truth. You will read here a part of the story that is usually unspoken. All the nasty things that you might expect or hope would be avoided in these institutions of higher learning: sorry they aren't. Intelligent, creative, analytical, think for yourself type people, like the professorate, like the practitioners of science, like those who are going to educate your children, are just ordinary humans. Imagine what dishonesty, theft of ideas and discoveries, hidden agendas, political ambitions, prejudices, profit motives, even sexuality can do to the education process. This is the setting in which the great wisdom of our world is transferred to the next generation. There is of course the other half of the equation as well, the students. What is it they want to learn? What is it they are capable of learning? What is important to them? Do they know what is important? As consumers will they accept someone else's idea of what is important? The full equation then has two main components, the producers and the consumers. Together they swirl about one another in a perpetual dance that in many ways is a competition. Both halves are driven by a profit motive. Neither half fully understands the other. In their entirety they form an unstable, dynamic environment, dangerous, challenging, rewarding, and not the least of all hilarious. Well, hilarious depending upon how you view it.

My dear partners please correct me if I am wrong but this is what I believe is NHM's take on higher education in our society and the institutions responsible for administrating it. He feels that we are not so much a country of universities as we are a country of un-universities, which is the term he gives to an institution with the following characteristics. At the perfect un-university there would be no classes, exams, standards, laboratories, or graduation requirements. In fact, no one would be graduated from an un-university. There would however be spirit, logos and stickers for car windows, parking garages, dormitories with television sets in every room, sports and athletic facilities including swimming pools, rock-climbing walls, skateboard parks, and either a golf course or an artificial golf building. Food halls would operate entirely on credit cards. They would include pizza joints, bars, and nightclubs, movie theaters and even tattoo parlors where either color or metal could be applied to skin. The campus would have grass and plants, shuttle buses, space for social and religious groups to meet, health clubs with tanning stations, as well as medical triage rooms. Cell phones, iPods and computers would be provided to each un-student. E-mail, Internet and the ability to download music would all be free and available everywhere within the WiFi campus zone. Depending upon demand there might be counselors and libraries. In some cases there might be bus routes to places where professors work on unknown things in locked buildings. If an un-student got the inkling to meet a professor rather than communicating via electronic means, an appointment could be made and the student would be transported to the appropriate building. After a period of residence at an un-university a certificate would be issued and the un-students could then move on to the real world where knowledge is a spun product sold for a purpose. A very small number, not more than ten percent of the un-students who showed some real talent for something or the possibility of thinking independently might be afforded a second chance, this time inside the locked buildings. The government would probably support these students. They would be prohibited from making any public or political statements, or organizing groups, and definitely not be allowed to vote. Un-universities would be carefully distributed throughout the country to assure that no person of the appropriate age could escape attending one. In this way the "democratic experiment" laid down by our founding fathers would be given the best chance to last forever. Open your eyes, NHM would say. This is what our universities have become and they are not similar to those in other countries. Could it be that given our population this is really all that might be hoped for? Could it be that others have made some perfect mistakes in building the educational system in this country? Creator, Performer, Guardian would you agree with my assessment of his perspective? Yes all four of us are in agreement on this. Readers, where do you stand? See if your position is the same when you have finished reading this history of a man who has been a professor for forty years, a man honored by inclusion in Who's Who Among America's Teachers, and a man who has seen first hand how European universities differ from those in America.

2. London is where NHM became one of the few who adds to what is known about the world

Even Creator, that brash somewhat arrogant fellow was amazed that we could have ever gotten from our origins to London. Performer didn't so much as pause to catch his breath, he just charged ahead. To me, Observer, London appeared as if a giant step, a lot to observe and learn. I wasn't quite sure if our legs would be long enough to negotiate the steep climb. NHM knew that it didn't matter. He felt confident that we could get to the top one way or the other. Even more than that, he knew we would get there. It was as if he could look into the future with some kind of superhuman powers. When he made good on this prediction there was no way any of us could bring him down to earth where he belonged in the realm of the mortals. In London, NHM made the transition from being a consumer of knowledge to being a producer of knowledge. That is not to diminish the value of Observer, who after all is the ultimate consumer of everything, but just to change the sources of his input. Creator working through Performer became the dominant input in London and forever after. All of us learned that as a producer NHM occupied an entirely different place in the sociology of science than he did as a consumer of knowledge. Others were now hungry to consume what he produced. Some became friendly or at least curious. Some unexpectedly became hostile and wanted nothing to do with him even before knowing what it was he knew, what he had discovered. Life was full of surprises. Creator was so busy assimilating all of this that had he been in another form he would have lost weight from all his activity. Someone meeting him for the first time then could have mistaken him for a hyperactive child.

This is what happened in London. On his first appearance in the laboratory NHM meet with Bill Hayes to discuss the research project he would pursue. Together they reviewed what had been promised to the National Science Foundation in the postdoctoral fellowship application and how that might fit into the progress that had been made since the time the proposal was written. Bill summarized what each member in the unit was currently doing and eventually came to a new project that he was very excited about. He told NHM that a small group had just been formed to work on finding the genes that regulate DNA replication and that this group was not using *E. coli* as the experimental organism but rather a spore-forming organism called *Bacillus subtilis*. No one in the MGRU had any experience with this organism and he admitted that those trying to get it going had made little progress using the standard methods they were familiar with based on years of experience with *E. coli*. Bill asked if NHM could help them get started since he knew that NHM had worked with *B. subtilis*. NHM told him that he would be happy to join their effort. Their goals, to use genetic approaches to determine how the replication of DNA was regulated, were right in line with an idea NHM had outlined during his graduate years. NHM had kept a book in which he wrote down issues worth pursuing once the technology was sufficient to investigate them, and the genetic/DNA project was one of them. Bill introduced NHM to the members of the group. They included Julian Gross, a former student of Bill's who had worked on gene exchange by conjugation in *E.coli*, Dimitri Karamata, a physicist who had worked on the development of the first electron microscope and actually built the instrument (now on display in the Science Museum in Geneva) that was later used in some very important experiments on bacterial viruses, and Michael Peacy, a technician trained as a microbiologist. NHM was assigned to share a laboratory with Dimitri, while Julian and Michael worked in another room. Soon thereafter they all settled down to hunting for the mutants that have defects in genes that control DNA replication. Before NHM arrived Dimitri had begun to isolate mutants of all kinds, hoping to later find the ones among them that were of interest by developing some kind of screening method to identify them. It was a labor-intensive approach. NHM devised an alternate strategy. He would try to develop a system that killed all the cells in a bacterial population that did not carry the mutant genes he was searching for. His plan seemed a good idea on paper but it was totally untested in reality. Creator knew that. He was a bit concerned too, but Performer wasn't put off by it at all. Experiments began and the routine of the experimental protocol took over. The same experiment was repeated day after day so as to obtain as many surviving cells as possible. Once into the routine, NHM observed something beyond what went on in his petri dishes. He discovered what the weather in London was like, as well as the way in which the scientists there pursued their work

When NHM arrived at the lab each morning the walk from either the White City or East Acton tube stations to Hammersmith Hospital, which was located between the two, was very cold and the Medical Research Council building was not very well heated. Few if any people where there at that hour. Those present were strangely lingering about in a 37° C room used to incubate bacterial cultures. They appeared to be examining their petri dishes in such detail that it took forever! Could they be defrosting in the growth room NHM wondered? That was it precisely. As it warmed up outside more people arrived for work. By mid morning things were humming but by then there wasn't much time before lunch so experiments that took longer to do were put-off until the afternoon. There were three places people could go for lunch: the hospital staff dining room, Bill's shack (not Bill Hayes but another man who had been rehabilitated at the Hospital after the Second World War and eventually given the opportunity to work nearby), or a local pub. One way to avoid having to decide where to go was to delay coming to work until after lunch! There were no time clocks punched at the MGRU and indeed some of the scientists working there adopted an afternoon-only work schedule. On his first day of work NHM was taken to eat at the staff dining room. One had to purchase a ticket for the meal before entering. The cost was minimal. The dining room itself was a large poorly lighted space. Everything appeared gray. There were long tables set with white cloths, plates and silverware. Women dressed in servant uniforms circulated about. They collected each person's ticket and served the meal. After several days of this NHM realized that all the food regardless of what it was called tasted the same, was the same color, and texture, and precisely the same lukewarm temperature. It was food to calm the agitated, precisely derived from that served in mental institutions. NHM's colleagues saw it more as an outcome of the poverty from the Second World War that had not yet been overcome.

One day a member of the MGRU suggested that it might be better to try Bill's shack rather than continue with it. There, for two and six (two shillings, six pence) you could get egg, bacon, tomato and chips with a cup of tea. True you had to eat at an outdoor table but there was a canvas awning that could be pulled down quickly when it rained. Still it seemed more appealing than the pessimistic staff dining room. And so the tradition of squeezing through a hole in the fence surrounding Hammersmith Hospital and going to Bill's shack for lunch nearly every day began. About every two to three weeks Bill sent what became known as a gut signal. Everybody got a stomachache after lunch. It meant go back to the staff dining room so Bill would have time to change the cooking oil. He knew that everyone would return as digestion permitted. On the days everyone went to a neighborhood pub, no further experiments could be done. The Bitter got the better of them. The loss of time did not appear to be of concern to NHM's colleagues. He thought better of it however and limited his participation in favor of pressing ahead with experiments. It was fortunate that he did. NHM had a finite time to get something done in London and his future depended upon it. His European colleagues were not so motivated. They appeared to him to be well adapted to their environment where research funds were very hard to come by, salaries were very low and could not be enhanced by working harder or longer, and survival meant coming to equilibrium with a very low energy budget. The best they could do was to power their minds. Their performers were very much less frequently called upon. None of them carried the overhead that NHM did and so he had to work in a different way than they did.

Late in the afternoon just before dusk the sun always appeared to break through the clouds over Hammersmith Hospital. It became very quiet in the MGRU at this time. Most of the people had already packed up and left for home before the inevitable fog and rain that followed. On his very first day of work NHM noticed that by the time he left for home there wasn't anyone else there. The workdays were indeed very short. NHM emerged from the MRC building and started off towards the East Acton tube station as if he had done that every day of his life. To get there he had to pass by an almost medieval prison, Wormwood Scrubs, which was located next to Hammersmith Hospital. A very tall wall surrounded the prison, perhaps 30 or so feet high. It had observation towers on top of it. Outside the wall behind the prison there was a large grass playing field that had to be crossed on the way to East Acton. One didn't need maps to negotiate it. When NHM was midway across it the fog settled rapidly and visibility was cut to almost zero. No lights or buildings, not even the prison wall could be seen. Not a problem for NHM who had night navigation training in the Army. But, after wandering about and not reaching the end of the field for what seemed like hours, he realized things were a bit spooky. A horn sounded to add to the illusion. It sent the shivers down his spine. Creator concluded it must have come from the prison and so Performer set the course toward it. Once there he circumnavigated along it to the front of the prison then set out along a street in the direction of the station. Streetlights were of little help. The station however was right there where it should have been. From that day on a new route was taken. It went in the opposite direction along a city street to the White City tube stop. The route passed over a bridge that crossed above railway tracks. Atop the bridge was a mechanical train signal, the kind with an arm that came down to indicate something to the speeding trains. It was precisely out of the old British movie in which a man was decapitated by an identical arm. Although there was no way this could happen on the way to White City unless you hung over the side of the bridge that is, NHM kept his distance, if for no other reason than deference to the movie. Well, perhaps just as a little extra precaution.

While grinding away on his project month after month NHM could see the way in which others pursued their work. It was not a high tech environment in the MGRU. The working tools were brains and standard bacteriological methods. Incubators were scarce. The central warm room served as a substitute for them. Autoclaves were scarce. In place of them each lab had one or two small pressure cookers that were almost constantly in use. Although there was a liquid scintillation counter for determining how much radioisotope had been incorporated into cells, such experiments were done only infrequently because of the high cost of isotopes. There was a single compound microscope, one or two analytical balances, and a few colony counters. These consisted of nothing more than a lighted box with a magnifying glass attached that could be used to either examine what had grown in a petri dish or to count the number that had grown. Most laboratories in the United States had so much more than this that it seemed miraculous that so much progress could have been made in the MGRU with virtually no instrumentation, and such short working hours. NHM's work there differed only in the amount of time he spent at it. Bill Hayes must have thought it excessive because one day he presented NHM with a newly hired technician, a man named Fahim Ahmed Kahn. Fahim had come to England from Pakistan where he had been the head of a small college. His goal was to change his passport from Pakistani to British and eventually to go to the U.S. for a Ph.D. Fahim was very bright, hard working, and talented in many ways. He came from a wealthy family and brought with him detailed recipes given to him by the family cook so that he would be able to prepare and eat the things he was accustomed to. He also carried to London a collection of what he called powders, the spices needed to make his dishes. Word got around London that he was cooking wondrous things and soon he was sought after to cook for weddings and other special occasions in the Pakistani community. One day he volunteered to make a special meal for all of us in our lab and we took him up on his offer. He got it all ready at his flat, then Dimitri transported him with all of it to NHM's flat where all the others had gathered. The inside of my mouth turned into smoking plasma at the first taste, that is how spicy hot it was. Everyone else ate as if they had no functioning taste buds, but Fahim was quite upset that his boss, NHM, almost needed a trip to the fire station. In the lab on the next day NHM inquired as to whether the others could taste flavors above burning. They all could, but that was not sufficient for Fahim. He insisted he do the whole thing all over again in a mild form that NHM could survive. His meal was again dragged to our flat, and it was delicious.

Fahim came in early and he went home late. He took responsibility when things went wrong. If the cultures were contaminated when he arrived in the morning he calmly took off his shoes and socks, sat down and waited for us to get in. Then he said he was ready for us to burn the soles of his feet as he deserved the punishment. We thought it a joke but it might not have been. Once he told us of his family's proud history as fierce fighters, and that he was fearless of war. This interested NHM in view of his recent military experiences. Dimitri asked him what he would do if he had to fight against an enemy tank. That is easy he said. I would strap on explosives and blow it up along with myself as an honor to my family. That was in 1966. Neither Dimitri nor NHM gave it a second thought then other than to view him as trying to build an image of himself as a hero. Today his comments must be viewed in a different light. We have learned that suicide bombers are precisely the tactic that Muslims have adopted in their war against western culture. At the time we ignored his boast and continued our good-natured dealings with him as a scientist. Nothing unusual transpired in the laboratory where Dimitri, NHM and Fahim did their experiments, with the exception of two incidents.

Dimitri, originally from Belgrade, moved with his family to Geneva during the Second World War where his father was a professor of mathematics at the University. During the year we worked together he had to make two trips to Switzerland on short notice, the first when he was notified that he had been granted citizenship, the second when his father died. Before leaving on the first trip he asked if there was anything he could bring us from Switzerland. NHM requested that he purchase two small gold coins that would be put away for his children. Such coins could not be purchased in the United States at that time. Fahim overheard this and asked if he too could purchase one, and Dimitri agreed. When the coins arrived, we were thrilled. Fahim examined his as if it were a mystical substance. During the weeks that followed Dimitri and NHM noticed that Fahim often took his coin from his pocket and stroked it. He apparently carried it with him at all times. While waiting for various steps of his experiment to be completed NHM asked Dimitri if he knew of the annual inventory of gold at Fort Knox in America. Dimitri said he knew that gold was stored there but not of the inventory. NHM then asked whether he had heard that each gold brick was picked up, recorded, and moved to another pile, after which the entire vault was sealed and vacuumed. Why do you think they vacuum it? I'm not sure he said. NHM went on, gold is very soft when as pure as it is in those bricks, and just touching it causes small flakes to come off, so the weight goes down and so does the value. Vacuuming collects all the flakes so that the total value remains constant. In the moment of silence that followed Fahim excused himself and ran down the hall to the analytical balance. Creator and Performer had worked their tweak on poor Fahim. Dimitri and NHM barely managed to stop laughing before he returned. In a much more serious incident Fahim informed us one day that his brother was coming for a visit to London and asked us whether we wanted him to bring anything for us. Dimitri couldn't think of anything but NHM did. He asked Fahim whether he could get a tiger-skin rug. NHM told him that he we would be happy to pay for it. Fahim assured him that his family had hunters and there would be no problem getting one. At about the time of the expected visit and delivery of the rug Fahim disappeared. He did not come to work. Neither NHM nor Dimitri could reach him at his home. There was no sign of him. He just disappeared.

Dimitri decided that he and NHM should go to his flat to see if anything bad had happened to him or if his neighbors knew anything about where he was. They drove there from the lab and after much pounding on the door managed to get a response from within. Fahim was there. They coaxed him to open the door a little. He looked miserable. Something had gone terribly wrong with his brother's trip. His brother was arrested at the airport in Pakistan for trying to take a tiger skin out of the country. Even though he claimed it was a prayer rug, the authorities took him to jail. Fahim's father, the family patriarch, was called and had to use all his powers to get his son released. Fahim was humiliated by not being able to keep his word and would not come out. Repeated visits by Dimitri eventually succeeded in getting him back to work. A few months after NHM returned to the United States Fahim was accepted into a Ph.D. program at the University of New Mexico, finished his graduate degree there and returned to Pakistan. NHM and Dimitri were never able to reach him again and to this day do not know whether he survived the many conflicts that have plagued Pakistan.

Dimitri too vanished one day. That was indeed unexpected. It turned out that his father was near death and his siblings had called for him to return immediately to Geneva. It was the evening and unfortunately Dimitri did not have the cash on hand needed for the journey. The banks were all closed and credit cards had not yet been invented. Dimitri somehow knew that Bill, the man at the food shack always had a supply of cash for his business and called him to ask if he could take a loan for this emergency. Bill immediately responded positively. Dimitri went to collect the money, then at the last minute asked what he could leave for collateral. Bill said it wasn't necessary, he trusted him. Dimitri asked what he would do if the plane he was taking to Geneva went down. How would anybody know that he owed Bill money? Bill replied that it would be his loss. Today that might not seem like much of a good will gesture but given the poverty that people like Bill faced in London in those days it was an outstanding act of generosity. Most working class people lived on the slimmest margin of solvency. They routinely overdrew their accounts at the end of each month then paid back the overdrafts from their next month's income. Dimitri departed and returned safely after the funeral. By the time he got back NHM had accumulated about twenty mutants that he believed might harbor defects in the genes they were hunting for. Dimitri had several hundred mutants but no method yet to screen them for those with DNA defects. NHM decided the time had come to characterize the ones he had isolated. He discovered that the analytical chemicals used in the standard DNA assay that he needed were not available in the MGRU. One of his MGRU colleagues however, Willie Donachie had an alternate method and he agreed to share some of the necessary reagents with him.

First NHM calibrated the alternate system. Then he began examining his isolates to see if any had defects in DNA replication. The final step of the assay involved boiling the samples in a solution that turned a color the intensity of which was proportional to the amount of DNA present. He coded all the samples with numbers blindly assigned to each sample so that there would be no bias in the initial assessment of the results. Dimitri happened to be there at the time the samples were boiling. Together they watched the color develop in amazement. It looked as if the very first mutant examined was what we were after. Dimitri took the specimens, each in its own little bottle, out of the beaker of water used to heat them and lined them up in what he believed would be their proper order. Indeed their coded numbers matched his order exactly. The differences in color strongly suggested that after this isolate had been transferred to the restrictive temperature it stopped making new DNA. We agreed however not to mention this to anyone until the results were checked and confirmed using an alternate method, namely a radioisotope incorporation method. By the end of the week NHM had confirmed his initial observation and it was time to let the word out. The first to be informed were the members of the MGRU. Rather than call them all together NHM decided to speak to his colleagues individually. That was not a good idea. The first person to whom he spoke immediately relayed the information to his technician. Observer will never forget how the technician responded to the news. "Oh no, not him. I mean why should he be the one to find the first DNA mutant?" he said. To which his supervisor responded, " Why did you say that? In Science we are all in this together." NHM knew at that moment, we were not all in this together. Good luck for one is bad luck for another.

Bill Hayes was pleased that NHM had found a gene that controls DNA replication. His own initial discovery of the first bacterial plasmid, F, stimulated ideas about how the control of DNA replication operates in cells that carry more than a single chromosome, in Bill's case the host genome plus a plasmid. A replicon theory was proposed suggesting that the start of replication would have to be regulated and thus there would have to be genes that govern the start process. Once started, replication was assumed to require an additional set of controls, hence other genes ought to be involved. A technical problem existed in finding these genes. To remain viable cells have to be able to replicate their DNA thus mutations that block replication are lethal. To get around this required the use of conditional mutants such as the one NHM had found. His mutant was a temperature-sensitive mutant. When his mutant grew at low temperatures it replicated DNA just as a normal cell would but if the temperature was elevated the mutant gene it harbored did not work and DNA replication stopped. The way it stopped was important. In NHM's mutant it appeared that only the start of DNA replication was blocked. Replication that had already started appeared to continue until the entire DNA molecule had been reproduced. In the parlance of replicon theory NHM's mutant had the properties of an "initiator" mutant. At the time the group in London was hunting for such genes in *B. subtilis* a parallel effort was underway at the Pasteur Institute in Paris using *E. coli*. Their approach was like Dimitri's: find as many conditional mutants as possible then sort through them later for genes of interest. In their collection they had identified some isolates with DNA replication defects, including at least one that behaved as NHM's mutant did, so NHM's discovery wasn't the first time anyone had ever seen such behavior. The Pasteur group however chose to focus on determining the spectrum of phenotypes represented in their collection rather than to fully characterizing the properties of any one of them. NHM chose to do the opposite. He set about determining the replication patterns of his isolate under various conditions where cultures were shifted from one temperature to another, and so forth. In a few weeks time, enough was known to confidently conclude that indeed the initiation of DNA replication was genetically controlled independently from other phases of replication. In the MGRU they called mutants that blocked only the start of DNA replication the "late turn off" class. Later when NHM returned to the States from London it was agreed that he would continue working with late turn off mutants whereas Dimitri, Julian and others who joined them would focus on genes governing other phases of the replication process. But Dimitri first had to find a way to identify the DNA defective mutants in his collection.

Dimitri realized that he could develop a screening method to help him find other DNA mutants using the properties of NHM's isolate known then as ts-134. Using a dual radioisotope incorporation protocol he took the behavior of ts-134 as a model for identifying initiator mutants. His method involved following the rates of both DNA replication and protein synthesis at the same time in mutants transferred from permissive to restrictive conditions. Eventually he found a number of mutants representing genes that control several aspects of the DNA replication process. Many of these fell into the "early turn off" category and could be easily distinguished from ts-134. Dimitri's mutant collection later proved to be a useful source for finding genes responsible for many other cellular processes concerned with cell growth, cell division and cell wall synthesis. In many ways his collection was equivalent to that produced in Paris at Pasteur first by Dr. M. Kohiyama, then later by Dr. Y. Hirota. A third collection produced by Dr. L. Hartwell in the United

States using yeast as the experimental system also became a valuable resource that was used for many years. Dimitri spent the rest of his career mining the mutant collection he produced during his years in London. NHM was very pleased that his prized discovery provided a tool that Dimitri could use not only to complete his PhD but to help him find the key genes controlling DNA replication in *B. subtilis* for which he later became famous.

When NHM was about midway through the characterization of ts-134 Bill Hayes came into his lab one day and asked if he would be willing to speak about his findings in advance of publication at a small meeting to be held in Cambridge. He said that an annual meeting had been organized the year before involving the members of the MGRU and those of the Laboratory of Molecular Biology (LMB) in Cambridge, a group that traced its origin to the original discovery of the double helix structure of DNA at the Cavendish Laboratory. For the first meeting, the Cambridge group came to London. Bill said it was our turn now to go to Cambridge. The format would be that five speakers from each group would present short progress reports about their research. The MGRU members were slated to speak before lunch, followed by the LMB people afterwards. A social gathering was scheduled for the evening at Francis Crick's house, a place adorned with a golden helix above the entry. How could any young geneticist turn down such an offer? When a copy of the program arrived, NHM saw that he would be the third speaker. In preparation he organized the results to be shown, had some slides made illustrating the results, and telephoned two colleagues, friends of his from the United States who were then doing postdocs in the Cambridge LMB, to discuss Sydney Brenner. Brenner had a reputation for attempting to destroy other people at meetings. It was something like a shark-feeding response that emerged as soon as he saw someone else's data. NHM knew he was particularly fond of feeding on young scientists just getting started as well as those who showed the slightest bit of insecurity about either themselves or their data. Brenner is precisely the kind of person NHM enjoys battling against. Creator and Performer would have drooled had they been able to just at the thought of getting a shot at him. Never mind that he was one of the originators of the replicon theory, co-director of a department at the LMB along with Crick, and on his home turf to boot. That did not faze them in the least. They thought it would be a wonderful challenge. NHM looked forward to being heard by the world leaders and showing them what he was able to do. In addition he couldn't wait to see if Brenner would bite on the simplest little lure that he was planning to cast in his direction during the talk. Dimitri and NHM drove to Cambridge together in Dimitri's old VW.

There were no disappointments in Cambridge. Brenner arrived late for the party at Crick's house and was true to form: both arrogant and belittling to those who bothered to speak to him. In the morning the first speaker from MGRU got up, gave his progress report, drew a meaningful dialogue from the audience, and Brenner did his best to beat up on him. Speaker number two was a Canadian geneticist Ken Sanderson; a person NHM knew well from his graduate days at Cornell when NHM was doing his undergraduate work. Ken was a Salmonella geneticist who had come to London to work with Bruce Stocker, a leader in that field, only to discover that his mentor was about to move to the States. When Stocker left Ken moved to the MGRU to finish out his project. Brenner could hardly wait to give his criticisms. Ken in his usual calm way thanked him for his suggestions and went on with business as usual. Finally it was NHM's turn. His subject was much closer to the interests of most of the Cambridge group than either of the previous talks had been. Ears were wide open to hear all the details. The audience seemed to feel very comfortable with what they heard, as if they had predicted such to be the case all along. When the Brenner-lure slide was launched the response was shocking. He shot up and argued with increasing agitation until he could hardly speak fast enough. Observer laughed so hard it brought tears to his eyes, figuratively speaking that is. Crick finally rose and said to Brenner, now Sydney just sit down and gather up your thoughts. If you want to ask a question later wait until the talk is finished. The next slide NHM had projected was meant to give the answer to what he assumed Brenner would become agitated about. Everyone laughed when they saw it. They realized immediately that the whole thing had been meant to bait Brenner, and it had worked perfectly. Creator believes it was the only time anyone had ever seen Brenner suckered like that. NHM couldn't help rubbing his nose in it. At the end of the talk he reminded them that what they had just seen was some of the strongest evidence at the moment in support of ideas about genetic control of DNA replication described in the replicon theory. These ideas had come from Brenner and some others in that very room. Brenner said not another word. That was in 1966.

Observer has found in his records an event that occurred some twenty years later involving Brenner that is worth inserting here because it shows that some personalities never change. In the 1980's professor NHM was back in Cambridge and went to see Brenner, who had become the head of the entire LMB. NHM was then a department head and wished to recruit a developmental geneticist to his department at the University of Arizona. It was a sunny warm day in Cambridge. Brenner and NHM went to lunch in the LMB dining hall on the top floor of the building. When he told Brenner the kind of person he was looking for Brenner shot back, yes, I have just the man for you. Performer let go causing deep laughter when he heard Brenner's suggestion. "Don't you remember", NHM asked him, "that when you were in Tucson this fellow had gotten some attention in the international press for his ideas, seemingly in conflict with your own and you told me what an ass he was?" If that is the best you have to offer thanks, please keep him, I will look elsewhere to fill the position. To which Brenner replied that he will have to think more about this. NHM doubted that he would. At that very moment Dr. Henry Mahler, a marvelous biochemist NHM knew when he was in graduate school appeared. He recognized NHM after all those years and asked if they could talk after lunch. What a pleasure it was to see him again and to find that he was still active in research at an international level. So far as the position at Arizona was concerned NHM filled it with a man who had been a postdoc at the LMB but not under Brenner's supervision. You will hear more about him when we get to the saga of life at the University of Arizona.

When Dimitri and NHM got back to London from Cambridge NHM settled into a period of intense work trying to get as much accomplished as he could before his year in London came to an end. Julian Gross appeared to focus his attention for the remainder of that time on mutator genes. He and his student, Patricia Hempstead had found an isolate that gave rise to many more mutants as it grew and produced progeny than the normal *B. subtilis* did. The mutants that arose appeared to be located randomly in any gene suggesting that the mutator strain might be unable to make accurate copies of its genes during replication. Julian's friend Lucien Caro arrived for a sabbatical visit from the States, and his student Claire Berg came along with him. Both of them brought their own projects with them although they seemed to be interested in the mutator project as well. Julian spent a lot of time with them discussing issues. Lucien Caro, originally French, wished to return to Europe so while he camped in London he traveled often to give lectures in search of a position. Lucien was widely respected for his work in molecular biology. Among his most significant accomplishments was the development of a quantitative whole cell autoradiography method. It provided the sole means at that time to study the distribution of various macromolecules in cells. For example, one could look in a microscope at a cell and see where it's DNA was situated. When Lucien discovered that NHM had found a DNA initiator mutant he volunteered to teach him how to use the autoradiography technique. That was the first time anyone had ever shown NHM how to do anything in research. Observer and Performer were thrilled. They went into action while Creator took a breather.

What NHM learned from Lucien turned out to be very valuable. Lucien insisted that they do the first experiment together. Performer actually did the work but with Lucien watching carefully over NHM's shoulder to make sure everything was done correctly. Part of the experimental protocol required that killed cells be kept in the dark so that a photographic emulsion applied over them would be exposed only by the decay of the radioactive isotope that had become incorporated into them, in this case into their DNA. Then the slides were developed in the same way as a photographic film. In the microscope the cells could be seen with black dots above them indicating the places where radioactive decay had occurred, showing in our case where the DNA was positioned. Together they put the slides, which bore NHM's mutant containing radioactive DNA, into the darkroom on a Friday afternoon. When NHM arrived on Monday morning Lucien was already there and he was smiling. Although the exposure time between Friday and Monday was really too short to expect to see many dots, he had anxiously developed one of the slides and examined it. What he saw was that the location of the DNA could easily be determined in the mutant cells making it possible to determine not only when DNA replication became turned off but where that happened in the cells and how the DNA's location was influenced by the continuing cell growth. Your system is ideal for my method Lucien said. As soon as NHM saw the results he realized that he would have to adopt this method as a main experimental approach. What luck, he thought. Not only did he learn the approach from the master but also in addition his experimental system was an ideal one for its application. Quantitative autoradiography served NHM well for years thereafter. Lucien got a professorship in Geneva, and Claire Berg eventually became a professor at the University of Connecticut. Lucien called NHM several years afterwards with an unexpected request. All of a sudden he could no longer make his own autoradiography method work. Is it still working for you, he asked? Yes it is, NHM told him. Could you send me the exact protocol you now use, he asked? Of course NHM sent it to him. When Lucien adopted the slight modifications of his own basic rules that NHM had developed everything started working again for him. NHM smiled with satisfaction having helped the master himself but neither of them ever knew why the modified method worked and the original one no longer did.

3. Travel had its ups and downs.

From London NHM traveled to Paris twice on scientific matters. The first trip was arranged so that he could meet with a professor from Yale who had heard about his work and wanted to see whether NHM might be someone they should recruit to Yale. NHM knew Dr. Adelberg only from his reputation. He was the author of the bacterial genetics portion of a very excellent book. Although NHM wasn't thinking much yet about jobs believing that having made a "great" discovery meant everyone would want to hire him, he thought it best to go and find out what the prospects were at Yale. So off he went to Paris. From there he continued on to the Government laboratory south of the City at the village of Gif-sur-Yvette. This government institute was housed on the grounds of an estate that had been confiscated from a Nazi sympathizer after the Second World War and converted to a microbiology research facility. A number of groups worked there including one directed by Professor Georges Cohen the man that Dr. Adelberg had gone to visit. NHM went by train, crossing the channel by ferry and then train again to Paris and Gif. Eventually he found his way to the Cohen group. Performer was not the least concerned by the fact that NHM's proficiency in French was rather marginal. He knew that everyone would speak English in the lab and that getting there was just a matter of maps and navigation. For that you don't have to speak to anyone. "Hell, this travelling is easy" NHM thought as he knocked on the door of Dr. Cohen's office.

From Dr. Adelberg, the visiting Yale professor, NHM heard all about the position there and the working environment. It didn't impress him. In many ways it appeared to match the undesirable aspects of his undergraduate Ivy League institution and he was certain he would not fit in well there. The same must have been sensed by Dr. Adelberg, for never another word was heard about the job opening or even, in fact, if there had been a job opening. During that visit a very unusual thing happened. At some point late in the day a group of about eight or ten people, NHM included, were gathered in Dr. Cohen's office. They were a mix of scientists, some French, others like NHM, from somewhere else, who were there for one reason or another. In a silent moment Cohen looked up at NHM and said to him, "All Jews". It caught NHM by surprise. Everyone remained silent. NHM glanced around the room at each person. And they looked back at him. Nothing more was said. Observer worked overtime in those few moments. NHM will never forget it. Creator immediately interpreted it as the pride of a World War II survivor marveling at the fact that others too had made it and eventually been able to achieve a place in the world based on their accomplishments. Unfortunately Georges Cohen who had moved to Institute Pasteur by the time NHM was a visiting professor there, was one of those in the audience who never forgave him for using the thumb pointer. To this day NHM does not know whether Cohen was one of those preyed upon by the others using it, or whether he just thought it inappropriate for a survivor to have done such a thing.

The second trip to Paris came at the very end of NHM's year abroad. He had gotten an invitation to speak at a small meeting in Naples to take place at the International Laboratory of Genetics and Biophysics. He and his wife decided that they should drive there making it a brief holiday as well as a business trip. They had purchased a small Saab in London to take back to the States and thought there would be time to go round-trip to Naples before departing. It would give them a chance to see parts of Europe for the first time. Once back in London they would release their car for transport and get themselves to South Hampton for boarding the S.S. United States and the journey back to New York. With two children under two they had to plan the logistics of their road trip carefully. The route would take them to Paris for a brief stop at Institute Pasteur. NHM had promised to give a copy of his initiator mutant to colleagues there interested in studying how it might influence sporulation, a process that occurs in *B. subtilis* but not in *E. coli*. Performer thought nothing of giving away NHM's prize even before he had published the first paper about it because he believed that all the key players in the field at least in Europe had already heard that NHM had found it. The woman who took charge of the culture, Dr. A. Ryter, was a distant relative of Dimitri's and had an outstanding reputation as an honest scientist. She introduced NHM to her colleagues and showed him around the Institute, a complex of buildings including a small hospital built initially for the treatment of patients with infectious diseases. There was a small museum in the residence where Pasteur had lived. NHM had no inkling at the time that he would eventually return to work there, become friendly with the curator of the museum and be given the chance to visit Pasteur's tomb. It was an incredible place beneath the residence adorned with mosaic depictions of the five-scientific/medical areas in which Pasteur had made key discoveries. After their brief stop at the Institute they did not fail to notice the beauty of Paris while hurrying through on the way south.

Although the Saab was new and drove well it was barely able to propel the enormous load it carried. In it were a month's supply of disposable diapers (nappies) to last the anticipated several week journey. All of the suitcases were on the roof along with two collapsible children's beds with mattresses and a British Comfy-folder pram. A propane stove, children's food, and a supply of two stroke motor oil that had to be mixed with gas in order for the tiny two-stroke engine to function were in the trunk. Maps, British, French, Swiss, and Italian currency, and the slides for NHM's talk were in the car along with NHM, Joan, Debbie, Marie, and the nappies. It looked like a circus on wheels. At one point along the way the travelling circus could have used a horse to pull it. In Switzerland on the way to Italy they decided to drive over the Simplon Pass rather than go through the tunnel as Creator suggested. What a mistake that was. With the little horsepower they had, they climbed and climbed in a line of cars, buses and trucks. Approaching the summit they reached a place where construction was in progress. It was stop and go for a while. A large bus led the way. Soon they reached a temporary detour that required ascent up a series of wooden planks to a platform above before starting the descent into Italy. The bus negotiated the planks then was forced to stop on the platform. The NHM circus had to stop just where the planks began. The line of vehicles behind them that stretched back a long way also had to stop. When the bus eventually continued, the circus wagon tried to start up the planks. The Saab had front wheel drive but without a rolling start it could not pull itself up at all. Several attempts to get going were made to no avail. There was no way to back up. Looking out the side windows it was straight down. Joan started to cry.

Performer took charge. He said, this is the best place to get stuck. Just watch and see. The route to Italy will not remain blocked for very long. As soon as everyone else behind them recognized their dilemma a group of construction workers came over to solve the problem. They picked up the car with everyone and everything in it and carried it up to the platform. Then they gave it a shove off the other end and it sailed without incident to the border and beyond. It passed through Florence and its famous museums, churches, and structures. They stopped in Rome, visited the Vatican, saw the Pope, as well as the Sistine Chapel, and discovered that the tires on the Comfy folder were no match for the cobblestone courtyards. Arriving in Naples a day before the meeting they went further south to find a place to stay. It was a beautiful small resort overlooking the Sea. At dinner some women from a very large Italian family seated nearby came to their table, picked up the twin babies and took them over to their own table. Twins were something special in southern Italy and their parents were viewed as some form of super-reproducers. The babies were passed around and around the table with much excitement and laughing. Neither Joan nor NHM could understand a word the Italians were saying. How do we get them back they wondered? Never mind, food eventually took precedence.

On the following day they drove back to Naples and checked into a hotel near the laboratory that had been recommended by the meeting organizers. It was just behind a very large soccer stadium that stood behind the laboratory. The hotel was a dump. They cooked on the floor of the bathroom using their propane stove. The Saab parked in a courtyard of the hotel became covered with pasta that other guests had thrown out their windows. The state of the laboratory was similarly awful. In this case it wasn't so much the buildings but rather the unfortunate scientists who had to work there. A communist-controlled union had seized power at the laboratory. The union members not the scientists or administrators came to determine whether or not each experiment would be done. Those scientists the union did not like were voted against, and the ostracized person could do no work. The dishwashers were in charge. No legal authority had the power to rectify the situation. At least two of the scientists had become mentally ill. One had to wonder who decided in the first place to hold a meeting there and what its purpose was supposed to be. Nevertheless the meeting began as scheduled. No sooner than two or three speakers into the program there came an ear shattering roar almost deafening in its intensity. Someone had scored a goal in the stadium. Each time the fans roared the speakers at the meeting had to stop and wait for the crowd to settle down before going on with their presentations. It was a good example of the fact that science cannot just be pursued anywhere without safeguards. Trying to work in an unsuitable environment is a losing battle. Unless you have the right local culture, there is little hope of any sustained progress. There is a cost that must be paid to provide an adequate setting for science and perhaps for all forms of scholarship. Here lies the problem: costs must be justified. Science that fails to pay its own way always lives or dies on the basis of decisions made by those in power. Even science that does pay for itself may fall prey to power. Add to that the fact that most scientists are employees and you see how uncertain the career of a scientist is. NHM did not dwell upon this however while heading back to London. Rather his focus was on time and money. Thanks to the disposal of diapers they could see more out of the car windows on the way back than they did coming. They drove as long as they could each day, avoiding mountain passes, expensive lodging and restaurants. Their estimates of how many miles per gallon they would get from the Saab and what the cost of petrol would be had both been way off. Fortunately they had sufficient funds to limp back to London.

They arrived in London on a typical rainy October day and found an inexpensive place to stay for the few remaining days. Although NHM's salary from his National Science Foundation fellowship, including dependent allowances, amounted to $ 7,500.00 dollars, which in those days put him in the top five percent income bracket in England, they had virtually run out of cash. During the year they were able to live in a pleasant flat located in the beautiful South Kensington neighborhood, and even to purchase the Saab from his salary, but the trip to Naples was much more costly than anticipated. So, they cut back as much as possible until departure day. Joan and NHM ate only one meal a day, taken at the central YMCA, a place where everyone short of cash went. The children had their normal meals and everything was fine until they hit the following snag. When NHM returned from his last visit to the lab Joan delivered the bad news. While she was out with the children doing last minute things one of them removed her shoe and gave it to a dog who then ran off with it. Before she knew it the dog was gone and so was the shoe. We had only one pair of shoes for each child. Although Joan hunted throughout the neighboring streets and alleys she never saw the dog again nor found the shoe. It was already dark when NHM arrived but he set out immediately to see if he would have any better luck. Unfortunately, he didn't. Not even the local derelicts had any suggestions as to where he might find the missing shoe. They thought he was crazy trying to find a child's shoe in that environment. Near closing time at the YMCA dining room he gave up the search and returned. Over dinner they counted all the money that was left. The next morning Joan located a place that had the least expensive children's shoes. Together they went to it, found a single pair of appropriate size, and purchased them. Upon examination the shoes were basically cardboard but that didn't matter. Their plan was not to allow them to be used for walking. They were worn more or less for show until they got to the boat. Once aboard, they took them off and put them away until just before docking in New York where they were to be met by waiting relatives. It worked. Both children had shoes. No one noticed that one child had perfectly normal ones the other paper-thin fakes.

On board they established themselves in their stateroom and went immediately to the dining room. Lunch was being served in all its grandeur before sailing and Joan and NHM broke their fast with multiple samples of each course all brought at once. There were parties in progress all around them consisting of guests visiting others who were about to depart on the voyage. The two of them focused on eating until they were surprised by the arrival of NHM's colleagues Willie Donachie and his wife Millie Masters, also scientists at the MGRU. There they stood, Champaign bottle in hand wishing us a bon voyage. Millie, an American met her husband at Princeton where they were both postdoctoral fellows in the lab headed by Professor Arthur Pardee, a famous bacterial geneticist. Her family was very upset that she had married a Scott and went off to live in Europe but it didn't diminish her career or the happiness of her life in any way. Together they were two of the most pleasant and smart people; creative and talented in so many things that it made NHM reconsider whether he shouldn't have stayed in Great Britain rather than return to the United States. Willie and Millie noticed that there was enough food on the table to feed an army so they dug right in without any inkling of why so much had been ordered. Soon it was time for the ship to leave the dock. Repeated announcements came over the speakers asking all guests to please go ashore. Their guests appeared to pay no attention to that. When word was broadcast that we were now leaving NHM became concerned that they might not be able to get off in time. They were the only guests left in the dining hall. He urged them to please leave or at least to tell someone that they had to leave before the ship was underway. Finally the boat began to rock as if moving from the peer. NHM jumped up and insisted they do something. Millie then convinced Willie that they really had to go, and off they went. Joan and NHM sat there almost positive they would reappear and that something would have to be done by the Harbor Police to get them off. The boat rocked twice more. People went to the windows to watch the ship's departure and to wave to friends or relatives who had come to see them off. Joan and NHM stayed put at the table too frightened to look. Five minutes, ten minutes, it seemed like an hour. Finally with no sign of them they started eating again and they stuffed themselves.

Good thing they did because within two days the ship ran into a fierce North Atlantic hurricane that caused many injuries to passengers and crew alike and seasickness that made it almost impossible to eat anything. Few passengers had the strength to go to the dining rooms. NHM, Joan and their children did but could hardly eat. The journey had been perfectly calm until the ship slammed into the storm. There was no warning. The initial blow was a rapid drop of the entire ship perhaps sixty feet that threw people up to the ceiling when it abruptly stopped and jerked upwards. Immediately thereafter crewmen rushed through the cabins and sealed all porthole windows with internal metal plates bolted to the wall. Within a day much of the crew was too sick to work. The storm continued for the next two days. The captain announced that arrival would be one half day late and that all injured parties would be taken off before others. Tugs met the ship outside New York harbor and pulled it in past the Statue of Liberty to a dock in south Manhattan. With the storm plates removed from the porthole in their cabin they could see ambulances waiting at dockside. On went the cardboard shoes. Each carrying one child, they headed towards the gangplank with their minimal luggage and NHM's collection of bacterial mutants. They then made their way to the customs hall and passport control. Once officially readmitted by virtue of a stamp in each of their passports, they moved on to customs and the first significant mistake of their journey back to America. This error cost them an additional four hours before they could pass from the holding area to the area where their relatives stood waiting to greet them. All because NHM had followed the instructions for importing bacterial cultures before he left England. He had obtained the needed license and had it with him along with the collection. Showing the license to the customs inspector led to their separation from the stream of people being cleared. They were taken to a separate location where a special agent from the United States Department of Agriculture came to officiate. He examined the license and the bacterial streaks, each in a little sealed bottle, then went off to try to determine the authenticity of the license. About ninety minutes later he returned, asked additional questions about what these mutants were and what they could be used for, and where NHM had gotten them, and where he was taking them, and whether they were dangerous, and on and on, all in a very friendly manner. NHM told him all about his project and that he was a scientist/professor doing research under the support of the National Science Foundation. The inspector was very impressed with it all. Then he excused himself and went off again, NHM assumed he went to get some verification from the NSF. When he finally reappeared he was very upbeat and NHM assumed they would soon be on their way. But it didn't end there. The inspector said to him, I know you from somewhere. NHM looked at him but even Observer couldn't place him. Could it have been in the Army perhaps? No it wasn't the inspector had served before NHM did. Was the inspector raised in New York where NHM had grown up. Yes he was a New Yorker but it didn't appear that they had crossed paths there. Was it perhaps in college? That's it he said, I'm Joe from entomology! Don't you remember we used to go on field trips together in my Jeep. That was it. They had both attended Cornell University at the same time, Joe studying entomology, while NHM was a botany and genetics major. NHM did often ride in his Jeep to laboratory exercises conducted in the fields or forests. Their reunion led to brief exchanges of what they had done for the past ten years. NHM thought that would have done it but no, off Joe ran to get his supervisor to whom he wished to introduce NHM. Joe gave his supervisor a lecture about the fantastic research NHM was doing, how they had been students together and on and on. He showed him the collection of strains and explained that it contained mutants that had never been seen before. The supervisor examined them so carefully NHM believed he thought you could see the gene defects by looking at the colonies that had grown on the little agar layer within each bottle. The boss asked many questions that required NHM to give them both a short course about what it takes to visualize the genetic defects. NHM got the impression that neither had much to do and that this was the most exciting thing to have recently come along in their jobs. Finally there were no people left in the large hall. It was dark outside. Even the longshoremen were gone. They were finally free to leave. Their poor relatives had no idea what had gone wrong. When the four of them finally got out, the waiting relatives grabbed up the children and they all left exhausted and hungry again. The difficult journey back was finally over. The cardboard shoes served their purpose. Home free all as they used to say in the game NHM played as a child on the streets of New York.

4. The professorial life is not exactly what you might have thought.

Getting established was not easy. Their Saab had to be retrieved from the docks, their household goods had to be taken out of storage and moved to the house they rented in Maryland, and they had to get there themselves. In addition, the American way of life had to be substituted for the British way of life. That took some doing. Once settled into their rented Maryland house they drove over to look at the new campus where NHM would be working. Just a few buildings had been completed at that time on the new campus but classes were already in progress and students were mulling about as they would on any campus. The setting was in fox-hunting country outside of Baltimore on grounds that had been part of a state mental hospital. NHM felt very much at home there. You see he had gotten his start in life living in a state mental hospital. That's another story though that we will get to later. On the grounds of this former hospital there were fields, streams and forested areas that were not yet transformed into what eventually became a full-fledged campus. Not far from this site there was a large parcel of National Forest that served as a greenbelt to the north and west of Baltimore proper. Small towns located nearby that dated to revolutionary times still retained their charm. NHM and his family eventually moved into one of them when they purchased their first home. Unfortunately they could not live in it very long. One of their children had become quite ill. They got the best advice they could from the experts at Johns Hopkins Hospital, the very place the children were born. The choices were not great. They could either adopt a very intense and undesirable medication regime the outcome of which was uncertain or they could move away to a warm dry climate. NHM spent some time evaluating the options, spoke to people at the National Institutes of Health, which was not far from where they lived, to get their opinions, and read everything pertinent in the biomedical literature. Having evaluated it all they elected to leave. NHM had been an Assistant Professor for barely three years there, had gotten his first research grants from the National Science Foundation, and had been promoted to Associate Professor at the time he had to move away. To achieve these things he had to live a whirlwind existence for those three years. Here are some of the things that went on during that compressed interval in their lives.

At Maryland NHM's career as a scientist and as a professor were both significantly advanced. He learned a lot about the lifestyle of these professions and had a good time as well. Did you know know that when you get a research grant from a federal agency they don't give it to you, they give it to your institution for your use? He didn't. Are you aware of the payments known as overhead that such institutions receive when a grant is awarded? He wasn't. Had he known about overhead he would have opted to compete for money from the National Institutes of Health rather than the National Science Foundation simply because the former provided the overhead funds in addition to those requested in your application, whereas the latter did not. For example, if NSF gave you $40,000 dollars, depending on the established overhead rate, the money you would actually have for your research project would be what ever was left after the university took its percentage. You might get $20,000! If NIH awarded you $40,000, the university's overhead would be added on top of that so you would end up being able to keep all you had planned to use. The idea behind overhead is to provide the host institution of a research project the money needed for operating the laboratory in which the project will be done. In other words if a research project requires space, then the scientist in charge of it has got to pay the equivalent of a rental fee for using the space. That's the theory anyway. In practice overhead money is spent on just about anything the university administrators wish to use it for. They might hire a professor of cake decorating using the funds that came along with a grant in support of bacterial genetics. At the new Maryland campus where NHM worked there was very little funded research thus the pool of overhead money available to the administration was quire small. Nevertheless it was well guarded. None of it ever came NHM's way during his tenure there.

In the 1960's, grant applicants were permitted to submit the same proposal to both the NSF and the NIH and thus to have two independent reviews made of the application and two decisions made as to whether it would be funded. That is no longer the case. Although the two agencies are totally independent they eventually began communicating with one another about proposals that were submitted to both. NHM had no idea who decided what about his own proposals but he was always notified first by NSF of an award and required to accept or decline it by a certain date. He discovered that he had to inform NSF of his decision before NIH made its notifications of grant awards. No one at NIH would tell him anything about his chances of getting an award before then. This meant that he always had to accept the NSF offer, and if NIH later made an award to him he had to turn it down. Sometimes an NIH official would call and ask if he had accepted the NSF grant. When he told them that he had accepted NSF's offer he was then asked to withdraw his NIH application so they wouldn't have to do anything further with it. NHM got the impression that he had become earmarked as a person who would only be allowed NSF funds, that NIH wasn't particularly interested in his research. It appeared that NSF was more interested in the scientist whereas NIH was more concerned with the scientist's project. Then unexpectedly one year the cycle was broken. NIH called and told him they would fund his proposal before even the NSF announcement date. He had no idea what brought this change about. Even though NHM had been previously supported by NIH as a predoctoral fellow he was somewhat hesitant to make the switch. NHM trusted the NSF to make good on what it said it would award him but he was uncertain whether NIH could be relied upon to carry through on their commitment beyond the first year of a grant. What should he do? He decided to call his NSF grant coordinator, Dr. H. Lewis, to ask his opinion. Lewis told him, "You have no choice but to take the NIH funds". He did, and wasn't disappointed. There were no problems dealing with the NIH bureaucracy. NHM continued nevertheless to submit dual applications at each grant renewal cycle for as long as it was allowed. This did lead to one awkward situation. After having turned down an offer of a grant award from the NSF in favor of another award from the NIH, NHM received a phone call from the scientific attaché of his US Senator congratulating him on the receipt of the NSF award. He thanked the caller for the Senator's interest but had to tell him he had refused the award! Several days later a letter arrived from the Senator, repeating the congratulations. There was no indication that he had any idea of what the circumstances were. The wheels of government spin regardless of details. Perhaps NHM should never have said a thing.

In the laboratory things went very well. NHM got a call one-day from a colleague at Johns Hopkins, Dr. Phil Hartman, who asked if his lab was running well enough yet to accept graduate students. He said, yes he could take one. Hartman told him that he had an applicant from the University of Maryland Medical/Dental complex in downtown Baltimore, who was well qualified in both microbiology and biochemistry, but that he couldn't take any more students at the time. Within a few days Bob Boylan was hard at work in NHM's lab along with an excellent technician, Sue Blough, who had been hired with NSF funds. Sue Blough became an expert on whole cell autoradiography and using it she carried out an elaborate series of experiments that NHM designed to characterize the way in which DNA segregates from parent to progeny cells when DNA synthesis is prevented. The findings revealed that cells of NHM's initiator mutant were able to grow and divide even though they couldn't replicate their genes at restrictive temperature. Cells arose that had no DNA. These cells were still alive but they could no longer grow and reproduce, akin to the behavior of human red blood cells. Word got out about this leading to a last minute invitation to speak at the national annual meeting of the American Society for Microbiology in Detroit in a symposium entitled, Regulation of DNA Replication and Cell Division. The organizer, Professor M. Schaechter, was well known for his work in this area. Together with Arthur Koch, an applied mathematician Schaechter published a classic paper that contained a quantitative model of the cell growth and division process. Koch was one of the speakers in the ASM session to which NHM was added. So was A.G. Marr, a man who, like Koch, used mathematical approaches to model cell growth behavior. When Marr completed his talk Schaechter asked him about the differences between his work and Koch's. Marr responded simply by saying that the equation Koch had written and upon which the Schaechter and Koch paper was based was wrong. It was simply wrong, he said. Schaechter asked Koch to respond. Koch got up and told Marr it couldn't be wrong because each of the terms in the equation had been generalized sufficiently to account for many possibilities. Upon hearing this, Creator awakened. What was the value of the equation if it could explain everything he pondered? How could one distinguish a correct mechanism from an incorrect one if the equation was so generalized, he asked himself. Observer watched Marr defend his position in a rather weak way. The issue was left up in the air then but Koch will reappear later as a player in the NHM story.

Following the sparring between Koch and Marr it came NHM's turn to speak. His topic was DNA segregation behavior when DNA synthesis was blocked at the start of each round of replication. He pointed out among other things that cell division went on in the absence of DNA replication. NHM didn't dwell upon it. It was not the focus of his talk. The audience however had not anticipated this finding. Most of the questions following his talk dealt with this uncoupling of the DNA replication cycle from the cell division cycle. Some days after the meeting a person who had been in the audience, Dr. John A. Gruneu, wrote to NHM as well as to the other speakers on the program in Detroit. Gruneu said that he would have liked to comment at the time but that he had an aversion to speaking in public. He informed us that his Ph.D. work in 1958 at Cambridge had dealt with the issue of linkage between the two cycles and in spite of growing evidence against his conclusions he stuck with them. NHM's findings appear to have shaken his confidence. He wrote, "5. Dr. Mendelson's Mutants. This one really hurts. …these mutants would seem to deal my general hypothesis the death blow. However…Perhaps in bacteria… Perhaps in Dr. Mendelson's mutants…. Perhaps……aw, nuts!" The dots are just as he wrote them, not things that have been left out in quoting him. From John Gruneu NHM learned that one person's discoveries can undo someone else's previous efforts. He took this as a warning that one had better not be too sure about things until others had solidly confirmed them. NHM came away from that meeting with the additional realization that his experimental system, well suited for studying the control of DNA replication, was also ideal for examining the relationships between central cell cycle events. The latter became a focus for him beyond just the behavior of DNA segregation.

Science is not, unfortunately, an enterprise open to all ideas and discoveries. Scientists are not all open-minded people willing to consider everything. Science is in fact controlled by a power structure that is parallel in many ways to those that dominate all other aspects of life. There are families and pedigrees in science just as there are in the Mafia and in Royalty. This power structure determines to a large degree who gets what job, which grants, which awards, invitations to meetings, editorial positions, appointments to government panels that evaluate grant applications, book contracts, and just about any other perk you can think of. Careers are governed by this power structure. Although the United States controls much of it because of its economic strength, versions of the same thing operate at least in the countries where NHM has worked, and most likely throughout the entire world. A very good mini-introduction to this structure came to him while he was working in Maryland. It all began with a phone call from a colleague who had gotten an invitation to attend, although not to speak, at a very prestigious meeting; the annual Cold Spring Harbor (CSH) Symposium meeting where, in 1968, the topic was, "Replication of DNA in Micro-organisms". The caller was surprised to see that NHM had not been included on the program in the session dealing with genetic regulation of DNA replication. He asked NHM if he had been invited to or knew about the meeting. The answer was, no, neither. Well, the caller suggested, why don't you call Dr. N. Sueoka, then a Professor at Princeton, who was acting as a spokesman for those working with *B. subtilis*. When NHM approached Dr. Sueoka he was passed off to Dr. John Cairns, then the head of the Cold Spring Harbor Laboratory, and in charge of the meeting. Cairns listened to what NHM had to say and made the following suggestion. He would have an invitation sent but NHM would have to show him his work when he got there and then Cairns would decide whether he would be allowed to talk about it. NHM agreed, had slides made, and gathered relevant data in support of that shown on the slides. He contacted relatives living not far from the CSH Laboratory on Long Island and arranged to stay with them for the week of the meeting. While standing in line to register NHM met by chance another attendee whose name he recognized from his nametag. It was Dr. H. Lewis, his grant coordinator from the NSF. He introduced himself to Dr. Lewis and thanked him for his help administering the NSF research grant that supported NHM's laboratory. Lewis was very receptive and said he was happy to meet NHM. Then NHM went to find John Cairns. They sat down together on the grass and NHM gave him the short course on what he had discoverd and wanted to present at the CSH meeting. Cairns looked over the data, was sufficiently impressed, and granted NHM ten minutes on the program but no more than four slides. OK, with just two publications to his credit NHM had gotten his foot in the door.

Only three hundred and fifty people had been allowed to attend the meeting. They were from all over the world, and they were clearly the power structure. A total of about ninety talks were to be given. Most of those attending held the best positions available anywhere. Many were highly rewarded with Nobel Prizes, and every other kind of honor. Those who hadn't yet gotten them acted as if they wanted them. They behaved like children clustering around their sports heroes, doing whatever they could to win favor. It wasn't just civil courtesy it was career building. Observer was flabbergasted. Creator was sickened watching those suckering up to Jim Watson, of the famous Watson and Crick model of DNA, a model that stimulated development of an entire field. Little did they know then how he and Francis Crick had obtained a critical part of what they needed to build their model from Rosalind

Franklin. Nor did they know who had helped them get her data, and how they used her intellectual property without her knowledge. NHM didn't know it then either but didn't have to. He had already reached his own conclusions about Watson's character. Watson revealed much of it in subsequent books that he wrote about himself and his science, as if it were something to be proud of. Although Watson and NHM had been to the same graduate school and had the same administrative head of their PhD committees, NHM discovered at CSH that the two had very different values. NHM saw nothing praiseworthy about Jim Watson. Watson wasn't the only eye opener however at the CSH meeting. In fact, he was merely a sideline in NHM's education about the people who dominated molecular biology. The main lesson came at his talk.

NHM was scheduled to speak in the session entitled, Genetics of Replication, which was the fifth of thirteen sessions that ran from Thursday the sixth of June through Wednesday the twelfth. That session began on a Saturday morning at 9:00 a.m. The first speaker was Julian Gross, NHM's colleague from the MGRU. Julian discussed the mutants that Dimitri had isolated and characterized, things that NHM had found about the continuation of cell division after cessation of replication, and the mutator strain that his student, P. Hempstead, had found. Both of these students were authors on Julian's paper but NHM was not. NHM did get mentioned in the acknowledgments as follows: "We gratefully acknowledge the collaboration of Drs. G. Bazill, Erela Ephrati-Elizur, N. Mendelson, and S. Zadrazil at various stages of this work." Before the session began NHM informed Julian that he was going to give the next talk following his. He asked Julian if he would like to show one of NHM's slides in his talk, having seen in his published abstract that he was going to say things about NHM's findings. Julian accepted the offer. Of course he had no idea that NHM had been limited to only four slides in his presentation and that by showing one for him it would permit more of NHM's data to be seen. Waiting his turn to speak NHM listened carefully to what Julian presented. When it was over NHM realized how important it is for young scientists to present their own work in order for them to become publically associated as an individual with their own work rather than only as a part of a larger group effort. Observer took careful note of how intellectual property was dealt with in the inner circle of molecular biology. It made him shudder.

When Julian finished speaking the chairman of the session announced, "we have an additional speaker now inserted into the program here, Dr. Neil Mendelson will speak on….." and the audience booed. Yes they booed although they had never heard of NHM, and had no idea of what he had to contribute. That moment itself defined the field and the profession for NHM. Creator did not miss the implications of this greeting. He warned immediately, do not go where you do not belong. These are not the kind of colleagues it will be pleasant to work with for an entire career. You and they do not have common goals. You do not understand science in the same way they do. They are not what you believe a scientist is, and although they are the leaders of the flock, it is difficult to distinguish who amongst them, if any, might be driven by the quest for knowledge rather than status and awards. Observer, thinking back to NHM's prior experiences and what NHM's core contained said, these are insignificant people. You really shouldn't waste any time with them. The Guardian quickly incorporated much of this new experience into the core and made it permanent. No one spoke however during NHM's short talk. They just sat there and swallowed hard as he showed them things they had never seen and interpreted them in ways that were beyond the scope of their thinking. On the way back to his seat he passed Dr. H. Lewis who was sitting on the isle. Lewis offered a big thumbs up as if to say they couldn't beat you down. Congratulations. You have been baptized under fire.

NHM sat through the rest of the meeting observing not only the science but also the sociology of science at work. He found that the vast majority of speakers had very poor communication skills. Many appeared to struggle with their work as if they had bitten off just a bit more than they could chew. Some began with apologies, seeking help from others in the room who knew more about their work than they did. Many gave few details in what they said but implied they knew much more. Whatever they knew they guarded well. Well-kept secrets appeared to be the goal of a scientific meeting! The social interactions outside of the lecture hall were informative as well. Dr. Julius Marmur, the man who developed the then standard method for extracting DNA joined a small group that NHM was speaking with and in deference to his stature everyone stopped talking. Although NHM only knew him slightly he had two startling things to say. The first was congratulations Neil you have made your contribution. NHM was shocked. His immediate thought was, is that all it takes. NHM thought this was the beginning of his career, not the highlight of his achievements. Maybe what Marmur had to say was a hint. Perhaps he meant that NHM should go do something else that would be more rewarding. Marmur's second comment was equally thought provoking. The conversation had gotten around to the work of H. Yoshikawa who claimed that multiple forks of DNA replication can travel down a molecule at the same time before the first wave has even reached the end of the molecule. Some people said they couldn't reproduce his findings. Marmur turned to NHM and asked, do you believe in Yoshikawa's stuff? NHM responded that he had no way to evaluate his claim because he hadn't done those kinds of experiments. Why do you question his results NHM asked? Because Yoshikawa isn't Jewish, he replied with a big grin. It was the second time in NHM's career that the issue of being Jewish had been raised and NHM wondered whether it would be a career factor. NHM was not raised in a religious household, nor did he live in ghetto-like communities. If anything, he thought of himself as a first generation secular American. These two wake-up calls and other events experienced while living in Europe taught him that it didn't matter what he thought, others saw things differently.

While all of this was swirling about at the CSH meeting a really brilliant man, Rolland Hotchkiss sat in the lecture hall throughout it all, just listening to what everyone had to say. He never took a note. When the meeting was over Hotchkiss wrote a summation of the entire thing. It was filled with details about almost everything that had been communicated. His article appeared as a summary beginning on page eight hundred and fifty seven at the end of the big red book, the Cold Spring Harbor Symposium on Quantitative Biology, volume XXXIII. In thirteen pages there he synthesized the state of knowledge of the entire field in 1968, including NHM's contributions! There in this prestigious book was NHM's third ever publication, pages three hundred thirteen to three hundred sixteen. Writing it was not pleasant however.

The format of the CSH meeting consisted of blocks of sessions starting each morning at 9:00 a.m., finishing before lunch, then starting again at 7:30 p.m. after dinner. The evening sessions often ran quite late. At the end of each day NHM headed back to his relatives' home where inevitably a discussion of the days events took place. To them this was an exciting insider's view of very important research, and they wanted to hear all about it. NHM attempted to convey some of the main ideas to them in language that a layman could understand and naturally they had many questions. It dragged on later and later each night as the meeting progressed. Pretty soon they felt as if they really knew something about all this hocus-pocus. They were almost ready to speak with authority to their circle of friends all of whom were in the business world. Another relative who came to visit them heard the word, "enzyme". "That fascinates me", he said. What fascinates you ?, NHM asked him. "Enzyme", he replied, "it just sounds right". Forget about what enzyme means. That was of no interest to him. Just something about the word in his mouth or his ear that's what mattered. It's hard to anticipate things like this but that's the real world. By the time all the eager listeners were worn out there was little time left for NHM to sleep. When he arrived for the first talks each morning he was dead tired. At the close of the meeting he headed home to Maryland totally exhausted. When he got there his wife and children were out shopping and the first thing he noticed was that the grass at his house badly needed cutting. They had a very fine manual lawn mower that was purchased when they bought their home, the idea being that the exercise he got using it would help keep him fit. Cutting the grass with it was indeed a workout. Tired as he was he thought that he better get started before the grass grew even taller, so he went to work. It was hot and humid and he felt lousy but he plugged away at it. When his wife returned she asked if he felt all right. Not really, he told her but he wanted to get it done. You don't look well she said. Why not stop now? The rest can be finished tomorrow. No, he thought he could finish it so he continued for a while but then realized that he had to stop, so he went in for a drink and to rest.

NHM fell asleep and awoke the next morning with a full-blown case of mumps. The childhood disease he never had before must have been transmitted to him from the neighborhood children before he left for the CSH meetings and he must have incubated it for the week he was there. NHM laid in a dark room barely able to move for ten days, but they were very busy days for him. The CSH laboratory of Quantitative Biology had its own press and it published the big red book, the Symposium Volumes, after each meeting. Speakers were required to submit their papers shortly after the meeting was over. He was worried that if he did not get his paper to them in time it wouldn't be included in the publication. So, he wrote the entire paper in his mind, every single word of it, while the mumps virus replicated in his body. He revised it several times until it was just what he wanted. Meanwhile he had his technician, Sue Blough, get the four figures ready for publication. They were taken from the four slides that he had been allowed to show at the meeting. On the first day that he could get out of bed he sat down and wrote out the final version, including the figure legends. To it was added the short list of six references, and it was mailed off along with the figures the next day, less than a week before the deadline. "Can Defective Segregation Prevent Initiation?" is what he titled it. The work that he had done although new and different from everyone else's fit into a defined field. It was part of a well-outlined discipline, and part of what was clearly regarded as the most important work of its day. NHM didn't know then but learned later in life how important it is for new discovery to be recognized as part of some existing knowledge. His hope then was that having a paper in this most prestigious book would be taken as a sign that he could be trusted to contribute and thus was worthy of further support.

Dimitri was not so lucky. At the time of the CSH meeting he had not yet finished his Ph.D. project and thus was subject to the hierarchy of scientific politics within the MGRU. This meant that although his work did get published in the 1968 CSH volume, he wasn't given the credit of senior authorship even though he had done the major part of what appears in the paper given by Julian. In the little group that NHM joined at the MGRU, Julian was considered the most senior. Although Dimitri was working toward a Ph.D. under Bill Hayes, he was not fully independent of Julian. When notice of the CSH meeting arrived at the MGRU, by then part of the newly formed Department of Molecular Biology at the University of Edinburgh not at Hammersmith in London any longer, Dimitri asked Julian if he could attend to present his findings. Julian would not hear anything of it. Julian went on to explain to Dimitri that no one attending the meetings was really interested in what was presented in the talks. The real value of the meeting was friends meeting one another and exchanging ideas. He told Dimitri that many of his friends would be attending, and they wanted to see him not hear Dimitri's talk. So, if only one person was going to be funded by the Medical Research Council (MRC) to attend it would be Julian not Dimitri. And indeed it was Julian who went off to Cold Spring Harbor. Dimitri was crushed by this introduction to the world of scientific politics. Long afterwards his wife told NHM that the frustration of it was so difficult for him to take that he actually cried as a result of the rebuff. The only time she had ever seen him cry. Dimitri's fate was far better however than that of many others: the deserving scientists who never were able to get an invitation to attend a CSH meeting and to have their work published in the big red book. They were understandably quite bitter about being excluded. Their ideas and discoveries were prevented from being exposed to the "in group" who controlled the discipline. For whatever reason the work of those denied admission wouldn't be categorized as part of the main stream, at the cutting edge, and among the best work. Dimitri at least got his name on a CSH paper. And he went on to a full productive career as a Professor at Lausanne, culminating with his hosting the meeting where the entire *Bacillus subtilis* genome sequence was assembled, and published. He has gone down in the record books as one of those who worked out the entire genetic code of this bacterium. The world is different now. The Internet has provided a way for everyone to record and pass on what they know whether or not someone else thinks it is worthy of consideration. The old adage from Plato's "Republic", "Those having torches will pass them on to others.", is now, " Those having torches can pass them to others regardless of what others think". The problem is, is there anyone left to receive them?

Although Bob Boylan, NHM's first graduate student, had only a brief period in which to work in NHM's Maryland lab, his progress was excellent. He started in a rather classical way looking for mutants that did one thing, then discovered one that did something else. In this case the something else was definitely worth a change in direction of his objectives. He discovered a gene that governs bacterial shape. His mutant could switch from rod to sphere and vice versa when given the proper signals. Boylan and NHM managed to get out a paper rather quickly about this without knowing that at the very same time Dimitri had gone as a postdoc to a lab in England where similar mutants had just been found although no publications had yet been written. Dimitri's mentor Howard Rogers lamented afterwards that his paper and the one from NHM's laboratory hadn't been published together, but neither of them at the time knew what the other had done. The Boylan/NHM paper led to several invitations and two significant collaborations. To begin with, Roger Cole, a very talented experimentalist interested in bacterial structure met them at a small meeting in Maryland. Cole was the head of a department at the NIH in Bethesda and he asked NHM if they could collaborate using his electron microscopy skills to examine the structure of the cell wall in our mutant. The work they did with him provided strong evidence that the defect leading to shape change must be a ramification of changes in the cell wall polymers. Then in the midst of it all came NHM's move to Arizona. Boylan decided he would come along to finish the experiments that NHM had agreed would constitute his Ph.D. project. In Tucson getting the new lab functional was not easy. The space was small and the available equipment was poor. Fortunately NHM brought along with him all of the core equipment that he had purchased on his NSF grants in Maryland. When he did get up and running it became clear that his little group was an island in a department in which very little research was being done. Money as well as equipment was scarce. So too were people working on anything related to what NHM did. The University of Arizona had just opened its newly founded Medical School and Hospital, located about a mile away from NHM's lab. In the Medical School were the only two people NHM could interact with scientifically, Harris Bernstein, a phage geneticist, and David Mount, a bacterial molecular biologist. Harris had come from the University of California at Davis where he was an Assistant

Professor, David from a postdoctoral position at UC Berkeley. Both Harris and David, believe it or not, had attended the 1968 CSH meeting, and Harris was even one of the speakers in the same session on Genetics of Replication in which NHM spoke. Unfortunately neither Harris nor David were knowledgeable in the research area into which Boylan's work drew us.

Bob Boylan's work did however progress nicely, aided by collaborators Roger Cole and another man, Frank Young. At that time Frank Young was working at the Scripps Clinic and Research Institute in LaJolla, California, a days drive from Tucson. His group studied *B. subtilis* cell wall polymers. NHM met him at a meeting before leaving Maryland. Young was anxious to work on the mutant Boylan had found and asked whether Boylan might consider doing a postdoc with him. Boylan was receptive to the idea. Final arrangements were made during Bob's stay in Arizona. NHM is sure Bob dreamed of the beautiful beach beneath the lab in LaJolla, but before he could get to it, Frank Young accepted the Chairmanship of Microbiology at the University of Rochester Medical School and that's where Bob had to go. The work NHM and Boylan did with Frank showed clearly which of the two major cell wall polymers was defective in their mutant. It was the teichoic acid rather than the strength-bearing wall polymer known as peptidoglycan. As NHM learned to his surprise at the first Gordon Conference meeting on the topic of Bacterial Cell Walls, the gene defective in the Boylan/Mendelson mutant was the first and only gene then known to regulate a bacterial cell wall polymer. Although NHM was an outsider in the field of bacterial cell walls, he found the people very receptive and encouraging. NHM maintained a relationship with many of them for the remainder of his research career. Most of the practitioners who worked on cell walls came from backgrounds in biochemistry, or chemistry. Their skills significantly overlapped with those who worked in molecular biology, the Cold Spring Harbor-type people. The difference between the two cohorts couldn't be overlooked however. What could be the reason for this? Why should one group be so cutthroat and obnoxious, the other so laid back and professional? Even Creator couldn't figure that one out. Could it be the difference between prize seekers and knowledge seekers? Could it be a founder's effect? Too bad those interested in the socialization of graduate science students couldn't have gotten into the 1968 CSH meeting, and for comparison attended the first Gordon Conference on Bacterial Cell Surfaces. It could have provided them material for many NSF grants, and papers.

Before we get to life at the University of Arizona there are a few things that must be mentioned about working at the University of Maryland, Baltimore County Campus. There, NHM had a lot more to do than just to wear the hat of a research scientist. His department head, Dr. Walter Konetzka gave it to him straight. Look he said, we are a four-man department in a new institution. All of biology is our responsibility. Every student you see out there represents money and justification for our jobs. We have got to quickly organize an up-to-date course structure that will satisfy the needs of those going on to graduate schools as well as those for whom their undergraduate degrees will be all the education they ever get. Those kids have got to be able to go out in the world and survive using what they get here. NHM got a sinking feeling at this point. How did he get into this? How could anyone find the time and energy to do anything at UMBC but teach? Is it too late to get out? He thought it was. Never mind he thought, I like teaching, But you didn't conceive of it as a full time profession the Guardian added. Well, what role can I play in this, NHM asked Konetzka? An important one he replied. You are the most broadly trained of any of us. Could you put together the first course they all will have to take in biology? Give them a really rock solid foundation hidden in the context of something they will enjoy. I know you well from your graduate days, he went on, I know what you can do. Oh God NHM thought mistake number two. In graduate school NHM had the reputation of being somewhat of a whiz kid, in part because a Nobel Prize winning member of his PhD committee had taken him as a tutor in molecular genetics. Professor H. J. Muller chose NHM because, as a member of his Ph.D. committee, he knew that NHM understood genetics and its history as well as everything being done in molecular genetics at the time. Konetzka, then a professor of bacteriology at Indiana University, was aware of this and now wanted to put NHM's knowledge to use in his new department at UMBC.

Fortunately NHM had about a month before his teaching duties were to start. He decided to organize a course around the theme of human biology. It would cover the basic areas of cell biology and physiology, developmental biology, genetics and infection and immunity using examples drawn from humans whenever possible. Doing it this way he thought he would be able to hold their attention, perhaps provide some actually useful information about such things as human reproduction, sex determination and its failures, genetic and infectious diseases, and other things that ought to be of interest to young adults. As Konetzka suggested NHM planned to hide in this format the core material identified as necessary in a national study dealing with undergraduate education in the biological sciences. Konetzka himself had served on the national commission and provided a compilation of topics that had been agreed upon as essential concepts. NHM worked around the clock to get as much of the material organized as possible before day one but there was no way that he could write or even completely outline every lecture. The semester ran sixteen weeks, classes on Monday, Wednesday and Friday, forty eight meetings in all. Not only that but to include all the students in each term, he had to repeat each lecture twice on the same day. NHM managed to complete only about a third of the lectures before the class started. The rest had to be done as the course went along. He fought to stay two weeks ahead. One of the most difficult parts was to get the graphics together in time. NHM wanted to have slides made to illustrate almost every idea. There were no computer programs then to do that so everything had to be done by photographers and graphic artists, and there was much demand for their time. The lecture hall was something that could give a professor nightmares. It was an indoor amphitheater like room that rose rapidly so that the last rows in the back were twenty or so feet above the small space in front where the lecturer stood. The whole thing was a poured concrete structure, no windows, no color; nothing comfortable, standard molded plywood seats with pull up writing pads. It was definitely a low cost structure. If you looked straight out toward the group when lecturing your view was directly into the crotch of anyone sitting in the third row. NHM's colleagues warned him about this given the scanty clothing being worn then. He thought they were joking but the joke was on him. The very day he gave a lecture on human reproductive physiology they attacked him. There in the dead center of the most revealing row, sat a group of girls wearing no underpants, wearing absolutely nothing at all under their clothing. At first NHM thought his mind was playing tricks, but he wandered about as he spoke and passed the show several times before he knew he was facing reality. He dare not let on or he risked losing his professorial credibility. He avoided making eye contact with them at all costs. Fifty five minutes seemed like hours. The subject matter at hand made it all the more difficult. Then he spotted his esteemed department head, Walt Konetzka sitting in the back. NHM assumed Konetzka wanted to be ready for any public outrage that might have been triggered by teaching these private matters so he attended to know for sure what actually went on in the lecture. The vast majority of students were not only very attentive throughout the class, but they demonstrated a kind of maturity NHM had not expected. That is probably what kept him going in the face of their incredible joke.

Finally the lecture was over. The students hurried off to their next obligation and NHM to his office. There he broke down into laughing so hard he failed to hear knocking on the door. The phone rang. It was the department office down the hall. Why don't you let Dr. Konetzka in they asked? What's going on in there? NHM opened the door and there he stood. He had no clue. He said he just wanted to let NHM know that he had never heard a more interesting lecture or one better suited to the kind of course being given. NHM told him what had happened. His reply: welcome to the academy. Then he too started laughing. It took the rest of the hour for NHM to settle down sufficiently to be able face the other half of the class. Upon entering the lecture hall this time he did two things. First, he moved a series of extra chairs that were in the room in case there was an overflow crowd or for those unable to climb to the regular seats, into a kind of barrier. The goal was to provide a few extra seconds should the group all of a sudden decide to rush forward. He realized it was a crazy thing to do but it became a habit from that day on until he left the University. An ounce of prevention he thought. How could he get out of a space like that? Military training surfaced. The second thing he did was that before starting to speak he took a careful survey of what he was going to have to face! No problems this time! Oh yes, just one additional thing. NHM noticed that the second lecture he gave was always a little bit shorter than the first. Because he didn't want the two halves of the class to get out of synch he always stopped at the same point in his lecture notes as he had reached in the first lecture. It must take less brain processing time to repeat the same talk after having given it just two hours before. On this day the second lecture was unusually short. The students had no idea why. NHM must have spent more time assessing the situation in the first lecture than he realized! Let's just put it that way.

There was a lot to be learned about student character in the teaching of a very large introductory course. In a class with hundreds of students you really get to see the full spectrum of personalities, abilities, and ethics. By and large the group NHM taught was anxious to learn and to make achievements. Many were looking far ahead, trying to build a solid foundation upon which the remainder of their education would rest and eventually from which their life plan could emerge. Although the Vietnam War was in full progress at the time, the students seemed optimistic and enthusiastic. The Nation might have been at war but the college freshmen were at school. In the first course NHM taught he tried to get to know the students a little better than would be the norm in a group this large. He experimented with an oral exam protocol in which groups of six came to his office and drew questions from a pool of cards laying face down on his desk. After taking a few minutes to get their answers organized, each in turn gave a little talk in response to the question that had been drawn. After each one finished giving his answer, the other students were asked to join in a discussion about the question and the answer. The original speaker then had to defend against criticisms that arose in the discussion. It was a pleasure to see how the better students dealt with these exams but it was painful to watch the weaker ones struggling. Those with poor communication skills were at a terrible disadvantage. Cheaters were exposed in front of their colleagues. It was very distasteful. NHM had to abandon the oral exam format. His colleagues felt vindicated when the oral exam format failed. They counseled him from the beginning to use only a multiple choice question protocol, pointing out that as class size increased each term there would be no other way to deal with the numbers. They were right. He began to write questions in conjunction with each lecture that he gave.

By the time of the first final exam NHM had several hundred questions to choose from. The University established an examination hall in a multipurpose building. There was place for four to six students to work at each table. A photo identification card was required for entry into the room and it had to be kept available throughout the course of the exam. When the exam was over he collected hundreds of answer sheets, each bearing a coded student number and the response to about seventy five questions. The main campus computer was given the job of grading all of them. He brought the pile of exams plus the answer code over to the computer room. There he found a young computer engineer who was working on a hard-wired board for a large Univac computer, about the size of an automobile, which stood in the middle of the room. The engineer took all of the answer sheets and said to call the next day. NHM stopped in the next day. The poor man was still there constructing the board. NHM asked if there was a problem. He said it was minor and that he would be ready soon to grade the papers. When, on the following day he was still unable to start the grading process NHM pointed out to him that the University required that course grades be awarded the following day and that he wouldn't be able to do so without the results of the final exam. Can we grade them by hand the engineer asked? NHM said he didn't think so. O.K. he replied, I will take them to the main campus at College Park and have them done there at the large computing complex. Off he went. When NHM arrived at his office the next morning he found a note saying that all was done and the exams were ready to be picked up. Sorry for the delay was written below. NHM examined the computer print out. He was amazed at the analysis that it provided. All the grades were plotted out forming a beautiful bell shaped curve with suggested scores that could be used to separate A from B from C grades. There were comments on each question revealing whether it was a good indicator that distinguished an A student from a B student or whatever. There were comments about whether a student who got question X correct was likely to also get question Y correct. All kinds of cross-correlations had been performed.

There were even comments pertaining to which class of student chose which of the other incorrect answers. And questions that appeared not to distinguish any particular class of student from another class were pointed out so that they could be eliminated from future exams if the instructor wanted to do that. Having seen all this NHM decided to have every exam that he gave at Maryland analyzed in the same way. By the time he left Maryland NHM had a collection of about one thousand questions and their corresponding analyses. He knew then that it was possible to construct an examination that would produce any grade distribution that you wanted. There was nothing magic about it. He found that disturbing. Was this an ethical way to teach and test? Fortunately he never again had to teach a large introductory course consequently he abandoned the format altogether. Only once in the following thirty five years did he have to write multiple-choice questions, and the outcome of that exam was a shocking lesson about his new institution. So far as the collection of analyzed questions is concerned they eventually became part of the NHM collected papers housed in the American Heritage Center in Wyoming.

The Baltimore County Campus of the University of Maryland eventually became an honors college with heavy minority enrollment. When founded it started on that route by developing a strong liberal arts curriculum. The sciences, humanities and arts were carefully woven into the fabric of the overall program. The faculty was recruited from everywhere, a collegial mix of scholars, teachers, researchers, and artists all willing to listen to one another and express their opinions on matters whether or not they pertained to their own pursuits. The students were mainly from the Baltimore/Washington area. They lived off campus. Many were older people, some already in the workplace, others returning to complete degrees started earlier, or to start again. Most of them however, were standard average American College freshmen. Student organizations soon emerged, as did literary publications and sports teams. In 1967 the UMBC Literary Magazine called "Dialogue" was formed. It offered an issue in its second year, Volume two, No. one, that featured, interspersed between poems and short essays, a collection of ten photographs showing a nude man and woman in dance like poses. The journal itself carried no title. The cover consisted only of one of the photographs. There were no references to the source of the photos in the issue, although it was later claimed that they had come from a gallery exhibition that had gotten excellent reviews. As might have been predicted the publication found its way to the local state senators, thence to Annapolis and the Maryland Senate. One of the outraged, Senator Harry J. Connolly, was quoted in the Baltimore Sun as having said, " My God it shows naked students wrasslin". Dr. Elkins, the president of the University, and Dr. Kuhn, the chancellor of the Baltimore-area operations were "invited" to testify before the State Senate about this matter and other issues said to indicate a "licentious trend (that) began a year ago". All except one of the Baltimore County State Senators were reported to consider the photographs suggestive, presumably suggestive of sexual relations among Maryland students. The issue boiled down to whether taxpayer's money had been spent, whether the University of Maryland endorsed the view that marriage, family and fidelity were things of the past, and whether there were no restrictions on male and female students visiting each other's dorm rooms. Naturally the senator who initiated the entire protest later claimed that the magazine "has aroused the protests of many parents and citizens indignant about the illustrations of nudes in suggestive poses", but offered nothing to support this position. The student editor of the Literary Magazine circulated an open letter to members of the university in which he strongly defended the right of his publication to support literary and artistic expression. His response was written at a level that made the Senators who escalated a trivial matter out of all proportion appear to be idiots. The president of the student government association also supported the position of the Literary Magazine. Finally the faculty was invited to participate in "Discussions of the Contemporary Arts", entitled "Sex In Art". Fortunately there were no further ramifications of this fiasco. NHM considered it a good example, and in a way a warning that what goes on at a University is subject to both scrutiny and criticism from those outside of the academic or intellectual community. It does not matter whether or not the critics are idiots. What matters is whether they have or can influence power to curtail academic freedom. Just as NHM has a Guardian of his core others have their own versions of the same thing. Whether the two can coexist depends upon what is in the core of each.

The profession of being a professor must therefore be inherently dangerous. Whatever a professor teaches may be construed as reflecting his core beliefs. If that is not congruent with someone else's beliefs, watch out for trouble. A nude mind is at least as dangerous as a nude body.

After the infamous fall 1967 issue of Literary Magazine, in the spring of that year the editors invited the University Community to submit artwork, writings, and other materials that they wished to have reviewed for publication. NHM had a small watercolor painting that he had done while in graduate school that he thought they might be interested in. It was a small black and gray image of a face dissolving into nothing, somewhat prophetic and pathetic at the same time. NHM knew this painting aroused a strong response. He once entered it, along with several other pieces of his work, in a street art festival. He called the painting, "Mankind's Fate". Although two of the other works he placed in the same show were liked and purchased, Mankind's Fate drew many criticisms from people who appeared to be religious fanatics. First they asked if he had painted it and if so did he really believe it. A yes and yes response then led to arguments to the contrary. Several offered him religious books and pamphlets. The intensity of their emotions was startling. NHM wondered whether the fact that these people were living in what some called; "the buckle of the Bible belt" in the mid-west had anything to do with their reactions. Mankind's Fate was submitted in April 1968 to UMBC's Literary Magazine. If it was accepted for publication NHM thought it would be a test of the waters in the Mid-Atlantic region but he never had a chance to measure it's impact. It wasn't because they did not accept it. Having heard nothing from the Magazine staff by late May NHM got in touch with them and asked that the painting be returned before the end of the spring term. Their response was minimal. By October he was concerned enough to speak with an attorney. On October 30th he sent a letter to the Magazine's faculty advisor informing him that if the painting was not returned to him by November fourth, legal action would be taken. The faculty advisor instructed the student editor to trace the whereabouts of the painting and to report his findings within four days. The fall edition of the Magazine would not be published until the issue of the missing painting had been resolved. At the appointed hour a letter was received from the editor. It said that the painting was missing. Whether lost or stolen was unknown. The University Administration asked if NHM had a copy of it. He didn't but thought he might be able to make one from a photo of the street show that had captured an image of it along with the rest of his paintings on display. NHM provided them the best copy he could make. They published the photo along with a caption that NHM wrote for it, in the student newspaper, The Retriever, on December ninth. It is reproduced below. The painting was never returned. Was it destroyed by someone strongly opposed to it? No one knows. But, NHM thinks it probably was.

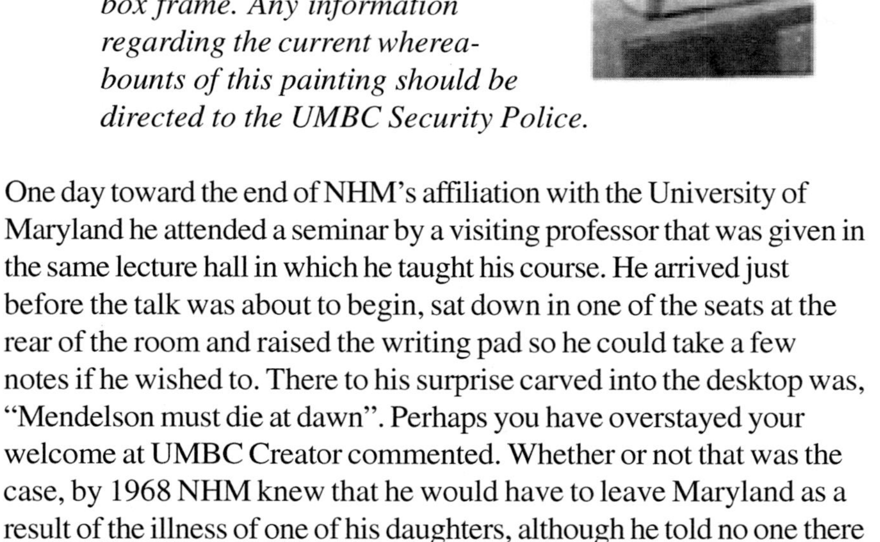
LOST OR STOLEN: The above *painting entitled, "Mankind's Fate" submitted by Dr. N. Mendelson to the literary journal DIALOGUE disappeared between June and November, 1968. The painting a grey watercolor, is about 41/2 by 7 inches and was mounted in a light-colored wormwood shadow-box frame. Any information regarding the current whereabounts of this painting should be directed to the UMBC Security Police.*

One day toward the end of NHM's affiliation with the University of Maryland he attended a seminar by a visiting professor that was given in the same lecture hall in which he taught his course. He arrived just before the talk was about to begin, sat down in one of the seats at the rear of the room and raised the writing pad so he could take a few notes if he wished to. There to his surprise carved into the desktop was, "Mendelson must die at dawn". Perhaps you have overstayed your welcome at UMBC Creator commented. Whether or not that was the case, by 1968 NHM knew that he would have to leave Maryland as a result of the illness of one of his daughters, although he told no one there about it. The main problem was how to find a job in the parts of the country that offered the best hope for an improvement in her health. Although NHM's research accomplishments were better known in the United States by that time, and a publisher had given him a contract to write a book based upon the course he had developed, he still lacked any real mentors, and had no network that could help get him a job. Desperate to move he decided to try writing to places whether or not they were looking for new faculty, places where he didn't know a soul and had no idea whether anyone there knew him. His three geographical choices were Arizona, California or Colorado in that order. He used the American Society for Microbiology listings of departments that offered the Ph.D. in Microbiology to find institutions that he would approach. He put together a file about himself describing his education and his teaching and research accomplishments, and sent it off along with a cover letter to the small group of institutions that he thought might be suitable. The first inquiry was sent to the University of Arizona in March of 1969. A copy of the cover letter is shown below.

Division of Biological Sciences
March 4, 1969

Dr. W. S. Jeter, Chairman
Department of Microbiology
University of Arizona
Tucson, Arizona

Dear Dr. Jeter:

In the hope of finding a more healthful environment for my children who have suffered from respiratory allergies since my arrival in Baltimore, I have decided to leave my present position. I would like very much to come to the south-west, particularly to a warm-dry region. The sort of position in which I am interested is one which will permit me to continue my current research and to participate in graduate or undergraduate teaching. I returned from an NSF Postdoctoral Fellowship in London to my current position at a new campus of the University of Maryland in 1967. In my current position I have had an opportunity to participate in the development of a modern Biology program at the undergraduate level. As part of that program I have organized a unique introductory course in Biological Sciences from which I am currently attempting to write a book. I am sending an outline of the general philosophy of the course.

At present approximately 40 % of my time is spent in research. My research concerns the control of DNA replication and segregation in bacteria. I am currently supported by a NSF research grant which has provided for a full time research

technician and nearly all of the research equipment required for my work. I am currently directing one PhD candidate, the first and only in Biological Sciences at UMBC, and one undergraduate honors research student. I have initiated a weekly seminar series covering the current literature in my area of research interests, and if I
remain at UMBC next fall, I will offer an advanced course in selected areas of microbial genetics and direct an undergraduate seminar in virology.

I have a broad background in biological sciences, starting in classical genetics at Cornell as a undergraduate in the Botany department and continuing into modern genetics at Indiana University in the Department of Zoology. My dissertation research was done in the Bacteriology department at Indiana University. I was fortunate and enjoyed a close relationship with H. J. Muller, Marcus Rhoades, Henry Mahler, Howard Rickenberg, Walter Konetzka and many other scholars present at Indiana during my graduate years. I am inclosing a curriculum vitae which fully outlines my background.

I am aware of the awkward manner in which I am attempting to locate a position in the south-west. Unfortunately, my research and other academic accomplishments are located in other geographical areas. Since my primary motivation for moving is related to the needs of my family, I am taking it upon myself to write directly to suitable institutions. I would appreciate any suggestions you may have regarding appropriate positions available in you institution or in neighboring institutions. I will be glad to forward letters of recommendation and any other further information which may be required. I will be giving a paper at the forthcoming American Society for Microbiology meetings in Miami and would be please to meet with you or a representative of your department if possible.

Thank you for your consideration of my request.

Yours truly,
Neil H. Mendelson

Looking at it now Observer realizes how unlikely it was that anybody could have found a job in this way. But against all odds on March eleventh NHM received he following letter from the Head of Microbiology at the University of Arizona.

THE UNIVERSITY OF ARIZONA

TUCSON, ARIZONA 85721

COLLEGE OF LIBERAL ARTS

[illegible]
[illegible]

March 11, 1969

Dr. Neil H. Mendelson
Division of Biological Sciences
University of Maryland, Baltimore Campus
5401 Wilkens Avenue
Baltimore, Maryland 21228

Dear Dr. Mendelson:

In response to your letter of 4 March 1969 I am pleased to inform you that we are currently seeking a person for our faculty with qualifications similar to your own.

We would like a person at the assistant professor level. Our salary range for the position is about $10,000-11,000 for the academic year (10 months) with supplementation at 2/9 of the academic salary by summer teaching or research support, making a total annual salary of approximately $12,500-13,500.

Our department is located in a building about 2 years old. We have approximately 40,000 square feet of teaching and research space, and I believe we are well equipped in both space and instruments for most types of microbiological research. We have presently 9 full-time and 5 part-time faculty members. Our gradute student population numbers 48, and we have about 200 undergraduate microbiology majors. The University of Arizona has 24,000 students this year.

Tucson is a city of some 325,000 inhabitants. It offers numerous cultural and athletic events. The climate, in our opinion, is unexcelled.

We would expect a person we hire to teach a course in microbial genetics for undergraduate and beginning graduate students (including microbiology and biological sciences majors).

In addition, we would want him to teach general virology (biochemical and physical characteristics of viruses) at the graduate level and participate in one of our large general microbiology courses as a member of a team effort. It is our intent that each faculty member should spend approximately 50% of his toal time in the teaching effort (contact, preparation, examination) and 50% in research. We would want our new faculty man to direct graduate student research as well as to develop a productive program of his own. We have close cooperation with the University-wide genetics committee and the Department of Biological Sciences.

I shall take the liberty of requesting letters from the persons whom you listed as references. If our situation sounds interesting to you, I would appreciate a statement about the research you would propose to do if you came here and an estimate of your space and equipmnt needs. If we reach a point in our negotiations at which it seems desirable, I would like to arrange for you to visit us, see our facilities and present your work before our group.

I shall look forward to your reply.

Cordially yours,

Wayburn S. Jeter
Professor and Head

By the end of April NHM made his first trip to Tucson. By the end of May he had arranged with the NSF to transfer his new research grant to the University of Arizona and he was ready to move. The University of Maryland informed him that he had been promoted to Associate Professor with tenure, but he had to decline the offer. Arizona agreed to change his appointment to the rank of Associate Professor, although tenure would be delayed for a year or two and the salary offered to him would not be increased. NHM was eager to accept anything Arizona offered to him. Although the move involved a significant salary reduction, a drastic reduction in the laboratory space assigned to him, and the need to purchase a house with much higher carrying costs than the home he left in Maryland, he did not think twice about accepting. Getting to Arizona was really all that mattered.

5. Go West young man if you have to.

The day they left their home in Maryland for life in Arizona was both a sad day and a hopeful one. Their beautiful little first home was someone else's now. All their possessions save what could be fitted into the Saab they brought back from London, were on the moving truck, and their older second hand Saab had been given to a company that would drive it to Tucson. In Tucson the house they had purchased stood empty but there was no one there to accept the shipment of their goods should it arrive before they did. The movers assured them that would be very unlikely, but also made it clear that if it did happen, their goods would go into storage and they would incur additional charges. At that time they had very little money. Their first home was purchased on a VA mortgage with no down payment required. They sold the house by themselves for several thousand dollars more than its purchase price but spent what was left after satisfying the lien for the down payment on their house in Tucson. Their savings were close to nothing, reminiscent of those last few days in London. The cost of the journey west had to come from the small retirement fund NHM had accumulated while employed by the University of Maryland. The Saab was heavily loaded again as in the trip from London to Naples and back. The journey was slow and uncomfortable. They did not have air conditioning in the car. The route they took required that they negotiate some high mountain passes, and travel across high as well as low desert. It was very hot and they were sick from the heat, the altitude, the dryness and the high light intensity. Nevertheless they pushed on. The anticipated arrival date in Tucson came and passed yet they were still days away. NHM questioned whether they were going to make it but finally a road sign read Tucson x miles. There was no turning back. Still they crept on at their snail's pace. They had to get there and eventually they did. Once off the interstate they went straight to a motel to sleep in a cool place. The next morning they made contact with their realtor. All their goods were in storage, and had been for several days. They picked up the keys to the house and went to look it over. It was located in a lovely community built around the ruins of an ancient Indian settlement that had become an archeological study site. It was a safe and beautiful place for small children and anyone else as well. The neighbors were very friendly. Many came with children in tow to say hello and introduce themselves. Some were professors, others physicians, some military officers, and others mid-career people of all pursuits. The previous owners had left them a set of instructions about how to keep the gardens watered, where the critical shut-off valves were for the major utilities and many other small details about the property. They were anxious to get started but first had to retrieve their furniture, buy some hats and sunglasses, and make sure they knew where to get medical help if they needed it for their children.

How would you like to have moved to a desert where people die from the heat and lack of water every summer and discover the day that you arrive that the house you purchased with your last dollar has no water? That's what they thought had happened to them. It was a horrible panicky feeling knowing that they had absolutely no way to buy their way out of trouble if they had gotten into it. Here is what happened. While their goods were being delivered NHM began the watering routine needed to keep the place green, the trees alive, and flowering plants in bloom. Most of the plantings were unknown to him. Some were poisonous others armed with dangerous stickers and thorns. He followed the gardening instructions carefully to make sure that he didn't kill everything immediately. When he finished the job he turned off the water and went inside to help unpack and set up the furniture. Later as dinnertime approached Joan discovered that water could not be drawn from the kitchen sink. NHM checked elsewhere in the house and found no water at all anywhere. He went to check the main shut-off valve behind the patio wall. It had not been turned off. Trying to keep calm, he then called the Water Company, a small private firm that owned the wells near the housing development and asked if there was a problem. They had no indication of any outage. Could we have purchased a dry house, he kept asking himself. Having found no solution he called back the Water Company and asked for help. They had no suggestions but agreed to send a person to assess the problem. This cannot be real, NHM thought but maybe we have been taken. He went back outside again this time to trace the route of water from the main valve to the house. He didn't feel very well though working in the heat outside so periodically he returned indoors to the air conditioning. When he regained his strength he went back out again on his hunt for water. Finally he found the problem and corrected it. At that very moment the doorbell rang. There stood two men from the Water Company. One asked if the water had come back on. No, NHM said, but I turned it on, and he took them to show them what had happened. A standpipe close to the house fed not only the house but also another faucet that fed one of the watering systems. When NHM shut down the watering he mistakenly turned off the upstream of two valves. The garden water went off as expected but so did the water supply to the house. The three of them laughed when he told them he thought he might have bought a house in the desert with no water. "No", said one of the men from the Water Company, "we have plenty of water here but it doesn't come from rain. We mine our water from deep wells; the water is five thousand years old," he said. Under the Tucson basin there appears to be a very large reservoir the total capacity of which is unknown. It's only very slowly replenished he added. But don't worry it will outlast all of us. They did not worry but someone else should have. The population growth in Tucson was so rapid that thirty years later Colorado River water had to be transported to Tucson via aqueducts to support development. Sure, they acted like big shots once the water was back on but the damage done by the scare was fully recorded by NHM's Observer. Things did not start off very well in Tucson.

The department at Arizona, to which NHM belonged, Microbiology and Medical Technology, was housed in an unusual building, known then as the Pharmacy-Microbiology building. Although it was not a particularly large building it housed the entire College of Pharmacy, as well as the small Microbiology Department. There were virtually no windows in the entire structure. Designed and constructed a few years before NHM arrived, the idea was to produce a facility within which infectious disease research could be conducted safely on the main campus. Such work would support the newly organized College of Medicine located about a mile away adjoining the University Hospital. The man responsible for the plan, Dr. Wertman, died before NHM arrived, and before the Medical School had adopted his program. In the end the Medical School wanted nothing to do with his idea opting instead to create its own microbiology department. Two of NHM's closest colleagues, Harris Bernstein and David Mount were among the initial professors appointed in the new Medical School department. The old Microbiology and Medical Technology (MMT) department was stuck in the College of Liberal Arts, had little in common with anything else there, and basically no overall mission. In an attempt to carve out a justification for itself the MMT department developed a number of service courses meant to meet the needs of pre-medical, pre-nursing, and other applied microbiology disciplines. When NHM arrived it appeared that the building was not used for the containment of any infectious disease work but rather it housed some undergraduate teaching laboratories, a few classrooms, faculty offices and conventional microbiology research laboratories. Though well lighted, the interior space was indeed cave-like. One could not tell what time of night or day it might be while working there. Those were the least of the problems.

Unfortunately the Pharmacy-Microbiology building, located near the southern edge of the campus is almost precisely under the main flight path approach to a very large military air base, Davis Monthan, situated at the southern end of Tucson. Davis Monthan Air Force Base was one of the most active Strategic Air Command bases in the country, second only to Da Nang during the Vietnam War. All kinds of pilots were trained at Davis Monthan and for good reason. The weather is ideal for flying all year round and the sparse population west of Tucson along the border with Mexico allowed for excellent bombing and gunnery ranges. In addition the dry mild climate makes it an ideal place to store surplus aircraft with little risk of deterioration. The predominant planes flying when NHM first arrived were F4C phantom fighter jets and U2 spy planes. The former, in groups of four abreast, passed over the Pharmacy-Microbiology building at low altitude all day long, sometimes with barely ten minutes between waves. The roar was deafening. No lecturer could be heard over it. One had to stop and wait for them to pass. It did not make sense that a major state University would allow such over-flights, but it did and does to this day. In recent years when base closures became necessary cost saving measures in the military budget, the State of Arizona did everything it could to prevent the loss or even the relocation of Davis Monthan. The current Governor, Janet Napolitano, worked hard in 2005 to try to block any future considerations of closing DM down. Davis Monthan simply brings too much money into the Tucson economy for any quality of life arguments to be heard above the roar of the aircraft and the dollars. What about safety?

Six years after coming to Arizona NHM had managed to move from the Pharmacy Microbiology Building to another building two blocks away, the Shantz Building, that at least had windows. It too was under the flight path. By then the main fighters passing overhead were the A7D aircraft. From NHM's office window he could see them approaching, hear them pass above, and then he could cross to his lab on the south side of the building and watch them progress towards the base. The jet engine sounds became embedded in his memory and that of everyone else's too who worked in the Shantz building. Late one afternoon the normal pattern of sound generated by the passing jets was different. As one plane flew over the building there was the sound of something exploding. This was followed a few moments later by a much larger explosion. But the roar of engines continued. At first NHM thought that a missile might have been launched accidentally. The professor working next door came running into his office. What was that he asked? NHM said he didn't know. Lets look out from the lab windows and see if we can tell what's going on NHM suggested. They both saw that the people on the street were running towards the south. Flames were rising from a schoolyard just beyond the UA campus. A plane had crashed. The first sound they had heard was the pilot ejecting as the plane passed over the Shantz building. The plane hit ground less than a block away. NHM left the Shantz building and rushed toward the wreck. He passed the pilot waiting behind Shantz for a helicopter to retrieve him. His plane had struck a car on the street near the school, killing the four girls in it, all UA students, before careening into the nearby playing field of the middle school. Burning debris was all over. Fireman and police where there in a matter of minutes. Later it was revealed that the aircraft had been having engine problems for much of its flight into Davis Monthan. The pilot nevertheless felt he could make it to the base, so he flew it over the campus as if it were the open desert, then he had to abandon it so he aimed the plane toward the playing field then ejected. He himself was unscathed. The dead below were counted as a cost of the cold war. Not an ideal location to be working in a collateral cost zone.

Back in the bowels of the Pharmacy Microbiology building NHM had been assigned a small laboratory space that included a room that you had to pass through to get access to a slightly larger space that was the main work area. The department provided him with an analytical balance. Fortunately the University of Maryland allowed him to take the equipment that he had purchased with funds from his NSF grant. Bob Boylan helped get the new laboratory functional. NHM hired a research technician, Linda Hallick, with funds from the new NSF grant. Although she knew nothing about either bacteria or genetics, Linda was a capable biologist, very bright and hard working. It did not take long for her to get up to speed. With the three of them working, there wasn't much room for anyone else in the lab. In spite of what Dr. Jeter had written in his initial letter to NHM, the department had very few resources. The emptiness of laboratories reminded him of an Edward Hopper painting. The lack of equipment in the department was reminiscent of the laboratory in London. There, with almost no tools, exceptional accomplishments were made. London was good training he thought for the real world. You never know what circumstances might bring. So he dug in for the long haul. And lived for thirty five years under the flight path. During that interval he occupied space in five buildings, some by choice others by circumstances beyond his control. Each move took its toll in time and energy. Although not quite as difficult as moving a cemetery, moving a research laboratory is not exactly like parking a Winnebago. It would have to be a large Winnebego at that given the fact that NHM never threw anything away. The four partners inside of him appeared to disregard the volume of things he accumulated even when it grew to unmanageable proportions.

Once NHM's productivity in the closet-laboratory became apparent it didn't take long for the department head to realize that he needed a larger place to work. There were graduate students who wanted to join his research group, and postdocs as well, but no place to put them. Eventually he was offered one of the regular laboratories, a room that had been occupied along with the two neighboring laboratories by a man who was working on tissue culture and viruses in sewage. Once the sludge was gone, they moved into it, and were soon joined by some new graduate students and a postdoc. John Reeve, had completed his

Ph.D. with Joe Clark at the University of British Columbia where he perfected the use of electronic cell counting and sizing instrumentation to study bacterial cell division mutants, and came to apply his expertise to the study of NHM's mutants. This objective could not be achieved however because a vital piece of equipment needed for it turned out to be a failure. Although a year was spent before John's arrival working with local engineers to build a state of the art instrument, the advice received from the consultants proved unsound. The resolution of the device built for NHM was inadequate. The project had to be abandoned. John's postdoctoral project was salvaged by a fortunate discovery NHM had made shortly after John arrived. A minicell-producing mutant was found that was similar to one previously discovered in *E. coli* by Dr. H. Adler. Very little was known at the time about such mutants. John was able to quickly shift his objectives, with the approval of NHM's grant fund coordinator, at NIH. NIH insisted however that NHM document precisely what went wrong with the engineering project that led to the failure of the instrument. NHM informed Dr. Jeter about this failure and asked him whether there was anything that could be done about the incompetent job done by the people at the UA College of Engineering. Nothing can be done he said it's the price of doing research. NHM was troubled by his response but too busy working on new experiments to take the time to pursue the dead instrument project. Pragmatism took over and it was a good thing that it did. Had he focused on infrastructure failures at the University of Arizona it would have totally consumed his efforts.

When NHM arrived in the MMT Department he realized that he was immersed in an alien culture and would have to be very careful to do nothing wrong that could jeopardize his position there. To do so he offered no suggestions and did his best to give the impression that: i. he saw nothing that was going on, ii. he agreed with whatever the group wanted, and iii. he was too busy with research and book writing to deal in any matters of scientific or institutional politics. This ignorant innocence could not last forever. The first department event that led to its breakdown came at the time a written Ph.D. qualifying exam had to be prepared for two students. Both students had completed all their course work and were then required to pass a written test covering the main areas of microbiology, not simply their areas of specialization. An examination committee was responsible for putting the test together, administering it, grading the papers and passing the results on to the department head. The head of the committee came to NHM and said that everyone on his committee felt that all of MMT students should at least know the rudiments of bacterial genetics. He acknowledged that the department did not offer any courses on the topic, and that there wouldn't be time for these two students to wait until NHM taught one, but nevertheless asked NHM to provide a series of multiple-choice questions covering what he thought they ought to know. That didn't sound exactly right to NHM but in conjunction with rule ii. above, he agreed to do so. A few days later NHM gave him twenty five questions that he believed would be fair, covering the kinds of basics that every microbiologist should know. The head of the committee thanked him and added that they might not use all of them but appreciated his having written them.

Following the doctrine that you should forget about what you have already finished and move on, NHM thought nothing more about the qualifying examination, that is until Bob Boylan came to see him. Boylan asked if NHM had written some questions for a graduate exam. Yes he did he told him. What about them? Well Boylan said, two graduate students had approached him and asked if he could tutor them because they had been told that some questions about bacterial genetics would be on their exam, and he agreed to do so. When he sat down with them they did not want to hear the kind of main ideas general review Boylan had prepared, but rather wanted to know specific things. Soon Boylan began to realize that they must have either seen the questions or been told about them. When Boylan described precisely what they wanted to know it was clear that something was rotten in the MMT department. The next morning NHM asked the head of the examination committee if he could make some minor modifications to the questions he had submitted, without saying why he wanted to. Sure, he said, take the entire exam and let me have the changes you wish to make. There were twenty of NHM's questions that had to be modified. What NHM did was to slightly change each one so that the correct answer was no longer the original choice but rather one of the other multiple choices.

After the exam had been taken, the answers to his questions were given to him for grading. Looking them over NHM felt as if he had just uncovered a big problem. He could not believe what he saw. Both of the students had chosen the original correct answer not the new correct answer to every one of the questions. NHM calculated the probability of that happening by chance and took all the evidence to the department head who looked it over and agreed right away, it was an open and closed case for cheating. Then he added, look these are both foreign students who will return to their countries when they finish here. I have taken charge of both of them because no one else in the department would. I know this institution and I tell you that bringing this matter up will result in nothing. Thank you for informing me of that NHM replied. I shall not press the matter any further, he added. NHM went home disgusted that he had ended up in a rotten institution, or at best just a rotten department. The Guardian of his core was badly hurt. But NHM needed this job and he stuck with it. Guardian said to him, you violated rule i. above, to which NHM responded, yes I did but I stuck with rule ii. And, I hope that saves me.

Professor NHM's second dose of reality about the University of Arizona came when he received a Research Career Development Award (RCDA) from the NIH. He was motivated to apply for this award for one reason only; it appeared to be a way that he might be able to enhance his salary. NHM accepted a low salary at Arizona when he made the move there without understanding that the raises given, if any, would be so small that he would never be able to climb out of a poverty situation. His $27,000 house was more than he could manage, his old Saab couldn't go over thirty five miles an hour, and his family's medical expenses were not met by the health insurance plan offered by the University. Two years into the new job it was quite clear that something had to be done to avoid the loss of NHM's home. The nepotism rules in place at the time meant that his wife, Joan, could not pursue any form of employment at the University of Arizona. With two degrees in genetics and many national awards during her student days there was no work for her either at the University or in Tucson. She immediately matriculated in the UA law school. NHM made application to NIH for an RCDA because it would provide five years worth of salary and reduce his teaching burden thereby permitting more time to be spent on research and survival at home. RCDA's were not meant to enhance an applicant's salary. He knew that. His hope was that because his UA salary was so low by national standards that NIH wouldn't realize that he was going to benefit financially from the award. NHM thought his chances of getting an award were reasonable considering those who had already received them. He recalls carrying the application papers from one administrator's office to another at UA in order to get the necessary signatures. The Coordinator of Research said, "Well that would be a nice feather in your cap if you could get it." And the dean of what was then the College of Liberal Arts, Dr. John Schaefer, said "Good luck" as if it were a joke to apply. Schaefer later became the president of the University, created a Vice-President for Research position, and transformed the institution from a teaching to a research university. When the notice of NHM's award arrived from NIH his wife was in class at the Law School in a neighboring building. He rushed over to tell her but just missed catching her before her class began. He read the notice letter several times to make sure he hadn't missed anything, paced around for an hour outside the lecture hall and hurried in to show her the good news as soon as the law professor finished his last word. We've got a little breathing room he told her. They both knew however that she would have to continue in law school and find work as long as they needed to stay in Tucson.

The RCDA award was received in a rather cold way in the MMT Department. There wasn't much enthusiasm for the lightening of NHM's teaching load, which meant an added burden for others. The Department had to find out how to put the award into effect. They sent NHM to see the Provost for details. While there NHM asked about the tax consequences of the award. The Provost had no idea. NHM asked him who had an RCDA in the Medical School that he might call for details. There was no one, he replied. Yours is the first one at the University of Arizona. No way, NHM thought. His impression was that all the strength in life sciences at the University was in the Medical School. It didn't make sense that no one there had yet received an RCDA. The issue of what will happen to NHM's University salary was raised. The Provost would not say anything about that. Therein was the heart of a significant problem. NHM's department head believed that the salary funds belonged to the MMT Department for him to use as he saw fit. NHM had heard from other RCDA recipients at other institutions that their salary money was returned to them for use in hiring a postdoctoral fellow. When no word was received for several weeks from the Central Administration regarding NHM's state-funded salary the department head called him in and said that he was going to contact NIH and turn down the award. Why, NHM asked him? Because the University is going to cheat me out of your salary if you get the NIH funds. NHM was infuriated by his logic but he did not reveal any of his anger. He knew instantly that he would have to violate all three rules he previously adopted for survival in the MMT department. Off he went to see the Provost again and told him that he did not want to lose the NIH funds. The Provost assured him, "We will not let him do that." A week or two later the money budgeted for NHM's salary by the University disappeared from the MMT department budget. Where it went was never learned. The department head was very angry. Although NHM tried to become invisible in the department, the anger spread to others in MMT. There was little NHM could do about it. Eventually he decided that it might help if he could move to another building but to do so he had to ask the department head's help. NHM approached the issue by telling him that he would really like to work in a building with windows and asked him if he should talk to someone in the central administration about finding some space. Oh, no the department head replied, I will look into it.

Later that spring the department head called NHM to his office and asked if he would consider moving to an old student laboratory in a building called Biosciences East. NHM asked if he could see the space before making a commitment. It was a very large room with place for twenty five students to sit at lab benches, tall windows stretching to a high ceiling on two of the four walls, and included a small adjoining room. There was adequate space along the other two walls to fit in large pieces of equipment such as incubators and centrifuges, but inadequate power to run them. NHM immediately accepted the offer, and requested that some of the overhead funds from his grant be used to make the needed modifications to the electrical service. Funding for the necessary modifications was agreed upon and the work soon was underway. NHM checked on progress periodically and happened by when a group of University electricians had just finished wiring twenty five new outlets on the bench tops and tripped the circuit breakers to power them. There followed twenty five small explosions. The electricians immediately shut the power down. "Oh well", one said, "I guess we wired them backwards", and they immediately began removing the burned out parts. Maybe it would have been better to stay quietly in the old windowless space Observer thought. Why don't you quit here and now Creator suggested. This institution doesn't have the infrastructure needed to support the kind of work that you are trying to do. Schaefer's idea of converting the University of Arizona into a research university might have been a mistake. NHM agreed. He could see no signs that the culture in Arizona could support the presence of such sophistication. This, "Great Desert University" as delusional members of the faculty liked to call it appeared to NHM to be as harsh an environment as the desert itself in which it is situated. Why didn't NHM quit? Did he know how to quit? Perhaps, but NHM couldn't leave the Tucson environment where the hoped for health benefits appeared to be working. He didn't have the financial cushion to go anywhere, and had no idea of how to get started in something else. NHM had dug himself into a deep hole in Arizona. There was no easy way out. In the end, it took thirty five years for him to admit how bad the situation had been.

When the electrical work finally was completed over they went into the old building with beautiful views and lots of space. A visiting professor, Dr. W. Segel from UC Davis, took a sabbatical leave to join them there for a year as a research associate. Soon after her departure NHM was forced to leave the space in Biosciences East. He never dreamed that when laboratory space there had been offered to him his department head had not gotten permission for NHM to use it from anyone with authority to assign its use. Perhaps the angry department head knew this would eventually happen and thought it might be a good way to get rid of NHM permanently. Here is how NHM found all of this out. Another professor working in the Biosciences East building asked NHM if he had seen a memorandum that listed reassignments for all the space in the building. No he hadn't. The colleague said he noticed that NHM's name was not listed in the distribution of the memo but that the person to whom NHM's lab would be given was included. Unlike all others in the building there was no mention of the space to which NHM would be moved. NHM took a copy of this memo to his department head who then told him that the building really belonged to the Ecologists not the Microbiologists, and that the space to which he had been assigned had really only been loaned to him. NHM went right to the point. I would rather not move back into the Pharmacy Microbiology building he said. The department head agreed to look for yet another space. A date had been set for the transfer of space in Biosciences East. NHM heard nothing about where he would be going so went to see the newly appointed Vice President for Research, Dr. A. R. Kassander. Although Dr. Kassander controlled no space, he was high in the central administration of the University and took up NHM's cause. Eventually he arranged for a group of rooms to be made available in another building, the Shantz building, and he agreed to provide funds to have the space gutted and redone to NHM's specifications. NHM designed a main lab and five smaller rooms to include an office. Everything was completed immediately before NHM departed for a year in Paris. As soon as the move to Shantz was over NHM organized a party to be held there to which he invited the dean of Arts and Sciences, the dean of Agriculture, and Vice President Kassander. Little did NHM realize that he would later have to move again out of this ideal space.

A scientist can gain some protection in a place like the University of Arizona by having a large amount of outside funding, and a good international reputation based upon achievements. NHM was fortunate to have both but it didn't help him keep his space in the Shantz building. That building was the property of the College of Agriculture. NHM was not in that college and those in the building strongly resented the fact that the best space had been made for some outsider who didn't belong there. On top of that there were two other sources of strong opposition to his being there. One was a professor who became angry as a result of events that occurred at a Ph.D.. final examination of one of his students. Another professor and NHM were members of the student's

Ph.D. committee. They discovered upon reading the student's thesis draft that none of the work done had been properly controlled. When the other faculty member raised this issue during the exam, NHM supported his position. Shortly thereafter the use of an autoclave in the Shantz building was denied NHM. Guess who controlled its use? We need it more now than in the past the disgruntled professor claimed. There followed the cleaning up of a walk-in cold room in which NHM kept some stored bacterial cultures. Sorry, there wouldn't be any space for you to use in it he was informed. OK, money protects against such things. NHM purchased a commercial double glass door refrigerator and stood it in the hallway outside his lab. Everything from the walk-in went into it. Of course the Fire-Marshall was called and ordered it out of the hallway. But it took two years to find a way for NHM to find space for it in one of his labs. To meet the autoclaving needs NHM returned to the use of large pressure cookers just like those that he had used in London. Life and research went on.

A much more serious incident arose however that forced NHM to consider whether to fight on or to just call it a day in Shantz. NHM had become a member of the Southern Arizona Symphony Orchestra and posted notices of their concerts in the hallway near his lab. He soon noticed that they were being torn down and thrown out. NHM asked those in his lab if they knew who was responsible. No one did but they agreed to watch out the next time. Just before the next concert NHM put up a notice. It was removed. Nobody saw the culprit. So, he posted two notices in its place. They too were removed. He posted four notices. This went on and on until there must have been fifty or more concert fliers plastered all over the place. Finally one of his students saw an Arab student from the floor above, tearing them up as if his life depended upon it. NHM had an Arab student in his lab at the time. She was an Algerian, educated in both Algeria and England before coming to the US to do a Ph.D. He asked her what the issue was. She told him that there was a very large group of religious fundamentalist Arabs attending the UA, and many of them were opposed to any western culture, including music. NHM found that hard to believe. About then a Mosque was built near the campus. The Imam was a graduate student in biochemistry. NHM had met him and found him perfectly comfortable with western culture. He developed a reputation for being a very liberal Muslim. Within the year the Imam was murdered in his Mosque, stabbed to death. Those responsible were never found.

Not long after that NHM received a call at home in the middle of the night. The caller informed him that there had been a flood in the Shantz building. Three floors were affected including the one where NHM's lab was located. He drove to the campus to assess the situation. NHM's main lab was totally flooded from above. All the drawers beneath the lab benches were filled to the brim with water. They had become so heavy it was difficult to open them. The flood came from the laboratory above, the very same lab where the angered Arab student worked. It arose because he had set a distillation system going then left without turning it off. All the reservoirs had filled and the water just flooded over and ran down to all floors below. A graduate student who had returned in the middle of the night to check on the progress of his experiments discovered the flood in the rooms beneath NHM's lab where he worked. NHM called the head of the laboratory above his and told him to come immediately to see what had happened. As sorry as he said he was, he did nothing to punish the student who caused all the damage. At the time of the flood there was in NHM's laboratory an instrument that had been designed and built by his collaborator at the University of Cambridge, John Thwaites. It was the sole instrument of its type in existence. Miraculously the instrument stood in a dry island surrounded by the soaked remains of everything else on the bench top, including all the specimens that had been made to be analyzed using the instrument. The State of Arizona reimbursed NHM for the cost of having all the specimens remade. Once the new specimens had been measured the instrument was sent back to Cambridge for safekeeping. The Arab had made his point. Unfortunately the US Government paid no attention to fundamentalist affairs in Tucson. Now it appears that one of the first Al Qaeda cells in this country was right here in our little town. Pilots trained in our little desert state. Even the FBI in Phoenix couldn't convince Washington that there was anything to be concerned about in Arizona. None of this is meant to imply however that the student who flooded NHM's laboratory was in any way associated with Al Qaeda or any other terrorist organization. Nor has it ever been proved that he purposely caused the flood responsible for ruining the specimens in NHM's laboratory. NHM believes it was no accident however.

The space that NHM occupied in the Shantz building was constructed for him during the end of his fourth year of support from the National Institutes of Health, Research Career Development Award. That same year the University of Arizona established a new policy to assure that the transition to a research focus would be successful. The performance of each department head would be assessed along with a review of the strengths and weaknesses of the department programs. The MMT department was among the first group of departments to undergo such scrutiny. A review committee was formed consisting of three professors from within the department and two others who were chosen by the dean from the faculty of other departments within the college. The MMT faculty elected the three so called, "internal members", a full, an associate, and an assistant professor. Unfortunately, NHM was the full professor they elected. That caught him totally off guard. It was the last thing he would have expected and he dreaded having anything to do with the evaluation of the department head. At our first meeting with the dean he told us that the full professor from MMT would serve as chairman of the committee and he outlined what we would be expected to accomplish. Yes, NHM had to chair the committee. Dread the thought. Later in private NHM asked the dean if he could be excused but the dean insisted that it was important for all departments to be reviewed in the same way, so he denied NHM's request. NHM was stuck with the assignment. The dean argued further that if NHM refused to remain on the committee it would be worse for him in the department than it had been. Consequently NHM assumed the responsibility and given the clear displeasure of the MMT department head with him, decided that the safest thing to do would be to conduct the most thorough review possible. Anything concluded in the final report to the dean would have to be based on very solid information. NHM drew up a plan that looked like a one hundred fold overkill and presented it to the committee at their next meeting. After much discussion they accepted it with minor modifications. NHM assigned each of the members a particular task to do in order to obtain the needed information. As a group they interviewed the key members of the central administration who had to deal with departmental matters. They solicited outside evaluations from other department heads both inside the institution and elsewhere. They asked for and received comments from national professional organizations dealing with the MMT profession. They sent surveys to all the former students who had been graduated from the undergraduate and graduate MMT programs. It was exhausting work and in the end NHM had the awful task of putting it all together into a report. The document turned out to be a small book, exactly one hundred pages long. By comparison other department evaluation reports were five to ten pages in length. In our report we concluded that the administration had no confidence in the head, that the head should be replaced and that the direction of the program should be altered. What a terrible thing NHM had to write about the man who hired him and made it possible for him to get his family relocated to the place where they needed to live for health reasons. When NHM handed the report to the dean he informed him that within the month he was leaving for a year to work as a visiting professor at the Pasteur Institute in Paris. He invited the dean to a party he planned to celebrate the opening of his new lab in the Shantz building, and asked him if he would deliver a copy of the report to the department head. His answer was no surprise. NHM delivered the document got on the plane and thought about the civilization that was waiting at the end of the journey.

6. An unexpected discovery takes over NHM's life.

The reason NHM went to Paris was not merely to enjoy food and culture. It was because he had discovered a totally new system and he needed a place where he could apply all his energies to getting it established. NHM had made other exciting discoveries before but this one really appealed to him. It was something fundamentally different from all the others and had the potential to open an entirely new approach in the analysis of biological form and function. It was as had always been the case for him in research, something totally unexpected. A year before being evicted from the Biosciences East building NHM's colleague, Frank Young, then head of the Microbiology Department at the University of Rochester Medical School called him. He said that he had heard NHM would be attending a meeting on the east coast during the summer and wondered if he and his family might like to spend a month or so living in Frank's house outside of Rochester, NY. Frank and his family were going to Portugal where he was to participate in teaching a course on laboratory techniques using *Bacillus subtilis*. NHM's wife, Joan, was almost finished with law school at the time and welcomed the chance to escape the heat of another Tucson summer so NHM accepted Frank's offer. Frank made arrangements not only for the use of his home but also for laboratory space where NHM could do some experiments while there. The Microbiology department was located in an old building that was part of the Strong Memorial Hospital affiliated with the Medical School. Frank's home was in the countryside just outside the small town of Pittsford. It was a pleasant setting and a nice change from Tucson summer. NHM brought bacterial strains with him and set about working to determine whether the walls of minicells turned over their polymers as normal cells did. The first experiment led to unexpected results. The bacteria did not grow in the expected normal way on the surface of the nutrient agar used in Frank's Department. At first NHM thought he had gotten everything contaminated but it wasn't so.

The unusual colonies that arose were filled with long helical filaments made of chains of cells rather than simple individual rod-shaped cells or short untwisted filaments. Within the helical structures there were minicells. Helical form must have had something to do either with the growth medium or the particular strain that NHM had brought with him from Tucson. Not wanting to be distracted from his original objective, he called his lab in Tucson and asked for the ingredients of the growth medium that they normally used, and also another strain of the minicell-producing mutant be sent to him. NHM put away the strain that produced helical forms to work on again after returning to Tucson at the end of the summer. Little did he realize then that he had found the experimental system that would consume virtually the rest of his career as an experimentalist. That was 1974, thirty three years ago. A lot has gone on between then and now. Frank rose to Dean, then Vice-President for Health Sciences at Rochester, escaping just before they were to terminate his administration position by getting himself appointed to be the Commissioner of the Food and Drug Administration in Washington, DC. From there he moved to the Department of Health and Human Services as a top resource for efforts in anti-bioterrorism. He attained Admiral's rank in the Public Health Service, and also became an ordained Presbyterian Minister. Is there anything more an MD/PhD could have done? NHM managed to stay one step ahead of decapitation at the University of Arizona, enjoyed discovery for the rest of his research days, as well as his life as a symphony musician. Tucson remained his home base, although he lived frequently in Europe. NHM and Joan always returned to Tucson however, and this is where they will probably die.

What became of NHM's discovery in Rochester consumed almost all of his intellectual abilities and almost all of his life as well. The swirl of events that he became trapped in reveals human nature at work in the world of discovery. As in any battle, people fought against his discovery and him, tried to steal it, and tried to prevent it from becoming too well accepted too quickly. Rushing ahead with understanding and breakthroughs in research is not well tolerated. Here is the way Observer recalls what went on. When NHM got back to Tucson after Rochester his labs were still located in the Biosciences East Building. There was no clue that he would soon be thrown out of there. The main focus was still minicell-producing mutants. Willie Segel, the visiting professor from California, a biochemist by training, was well into her project when NHM showed her the helical structures that he had found in Rochester. She realized a lot could be done with this new multicellular system and volunteered to see if the cell shape change could be corrected using biochemical approaches. Willie and NHM were the only two in his group working on helical structures. Growing them daily they soon learned how to obtain better and better structures. Starting from spores NHM was able to produce structures of amazing regularity as shown below.

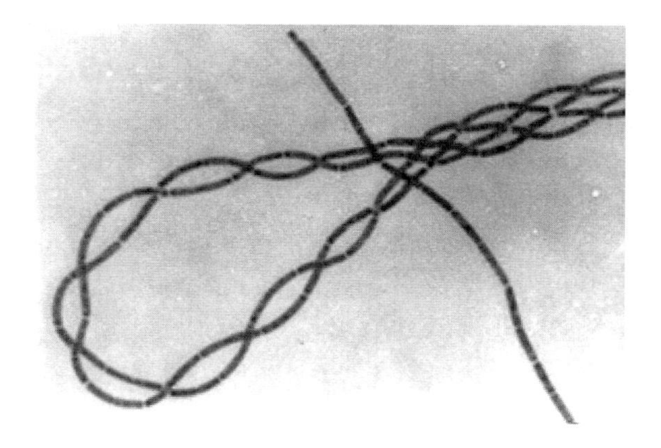

These beautiful structures could not be ignored. NHM got the idea that because of their regularity he might be able to find the rules that govern how structure passes as a phenotype from parent to progeny cells during growth and division. While grappling with the problem he learned that he was going to lose his laboratory space. Even so, he was consumed with finding the answer to the puzzle of helical form and its passage to progeny. He worked at home building models and challenging them with all the known facts about how bacterial cell walls are assembled at the molecular level. As he got closer and closer to the solution he seemed not to care what went on at the University. He had a strong conviction that he could crack this problem and he did. The shape seen resulted from a deformation caused by twisting with elongation and an impediment to the twisting motion. It was a whole new idea about the geometry of bacterial cell growth. All the pieces fit together but NHM realized that he was alone in this. He felt as if he had found a secret and was the only person alive who knew it. He decided that he had better write it all down quickly in case he suddenly died. He tried explaining it all to his wife and a few colleagues who he felt might understand the basic idea but that was no substitute for publishing it as quickly as he could. He worked day and night until he finished the manuscript then asked a senior colleague, Dr. Herb Carter acting head of biochemistry if he would communicate it for him to the Proceedings of the National Academy of Sciences.(PNAS) Carter agreed to have it reviewed. He asked if NHM could provide him with the names of some people to whom he could send it for evaluation. NHM told him he would be happy to, but he couldn't think of a single person at that moment who would be an ideal reviewer. NHM realized that something as totally new and different as this was didn't fit into the categories of existing areas of knowledge, or the disciplines that people consider themselves to be experts in. This wasn't biochemistry, or genetics, or microbiology or cell biology or bacterial physiology, or molecular biology. It was a strange mix of geometry, physics, and mechanics, dealing with growth and form. Never mind that it might be the way cells work, if it was going to get published it had to fit in with the way reviewers work. In the following few days NHM made and re-made short lists of potential reviewers. In the end he presented three names to Dr. Carter, all of whom were microbial biochemists who knew him and valued his previous research findings. Of the three Carter said he recognized one name and told NHM that he thought highly of this individual's work. Two reviews were eventually returned to NHM. Both were favorable and both to his surprise had the same comment: "this was going to be a very controversial paper". In spite of the warning Carter agreed to communicate it. It was the first of four papers that NHM eventually had published in the prestigious PNAS journal. He recalls thinking at the time, if the model described in his paper holds true, perhaps they will eventually elect him into the National Academy. It did. They didn't. So much for scientific politics and NHM's conception of the way in which the world works.

After NHM's paper on helical growth of *Bacillus subtilis* had been accepted by the editor of PNAS but had not yet appeared in print he was invited to give a talk at the annual national meeting of the American Society for Microbiology. These were generally large meetings sometimes with attendance in excess of ten thousand people. He knew the session to which he was assigned would draw a large audience. NHM thought it would be a tremendous opportunity to present his new discovery to them. There were of course two things that he would have to tell them: first, the discovery itself and then his interpretation of how cells that normally grow in cylindrical form become changed into helical form. It was going to be the most enjoyable research presentation he had ever given, perhaps the most enjoyable he would ever give. NHM could hardly wait to get to the lectern. The room was indeed jammed, with people standing at the rear. The chairman of the session called for a short break after the speaker before NHM finished his talk. NHM used that time to check the order of slides that he was going to show and to get the slide projector focused on the first slide. People were streaming back into the room, taking their seats and finding places to stand where they could see well as he fine-tuned the focus. The image on the screen was that of the beautiful double strand helix that had grown from a single spore. It was Figure one in NHM's PNAS paper that they had not yet seen. NHM waited for everyone to get settled, looked out at the crowd and said, "Need I say any more?" The audience laughed. He thought, quite a different reception than at Cold Spring Harbor seven years earlier. Then he answered his own rhetorical question by saying, "Yes I think I do because the helical shape you see isn't what you think it is." My goal here, he told them is first to tell you a little bit about this structure, how I found it and what I know about it, then to explain why I believe it isn't what it appears to be. You see it is not just a mutant that produces an abnormal rigid cell wall in the form of a helix, but rather it is a shape produced by deformation of the cylinder in normal cells. Motions linked to growth are at the heart of it. It's physics and mechanics not pure biochemistry. He told them the full details of what he was going to say at this meeting would soon appear in a PNAS paper. And he went through it all as quickly as he could to leave time for what he knew would be many questions. When he finished he had the impression that the audience realized that they had heard something very new, revolutionary in a way, and they wanted to get as much more detail as they could.

Midway through the discussion period, NHM asked one of the senior contributors in the audience, Dr. R. G. E. Murray, a man who had been the president of the American Society for Microbiology, and whose father had also been president, if he would like to make any comments. Murray rose, and with no warning whatever delivered a scathing tirade about how it couldn't be true. His argument went further and further off course until he ended up saying something about men who don't eat meat being inferior, with women, to those who do. Perhaps he thought NHM was a vegetarian? NHM responded to him by only saying that he thought it would be prudent for him to wait until the dust settled before coming to any firm conclusions about his model. Then a near riot broke out. Others in the audience refuted Murray's comments although they themselves didn't yet know enough to do so with correct arguments. Then Murray's supporters tried to defend him. The session chairman couldn't control the argument. Finally NHM said, well thanks for inviting me to speak at this meeting, and went to find a place where he could listen to the next speaker, a person he knew quite well, Dr. H. E. Kubitschek. When the room finally quieted down Herb said, "Well it is always hard to speak following Neil, but this is about the hardest it's ever been. What I have to say won't be quite as exciting."

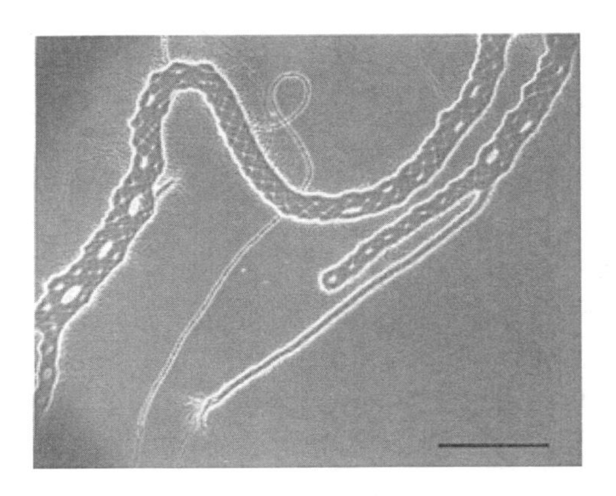

Within two years, Murray, then the Editor of Microbiological Reviews, and NHM met at a meeting in Halifax, Nova Scotia. At this meeting NHM reviewed his model of helical growth and showed as well the newly discovered macrofibers that could be produced by helical growth. This time Murray asked NHM if he would write a review about the helix system for the journal he edited. NHM agreed to write one. Dr. Arthur B. Pardee was also at the Halifax meeting. He and NHM knew one another from the days when NHM worked on DNA and cell cycle mutants. NHM asked Pardee at dinner the night after his talk on macrofibers if he would communicate a paper to PNAS for him about them. Pardee said yes; no one had ever seen them and he would definitely consider having a manuscript reviewed and sending it for publication if the reviews were favorable. The paper that he communicated was NHM's second publication in PNAS. The next time

NHM saw Art Pardee, Pardee told him that after his paper appeared in the journal he met, " a very famous scientist" on the street in Cambridge, MA who congratulated him on having communicated the paper. NHM was very glad to hear that for two reasons. First he knew that Art was very nervous about communicating papers in general and NHM was glad that he was reassured in this way that he had done the right thing. NHM was of course also pleased that his work was highly regarded by someone considered distinguished in our area of research. Art never did reveal who the "famous" man was. Could he have just made that up?

Before leaving Paris to speak at the meeting in Halifax NHM realized that there was a man on the faculty at Dahlousie University, where the meeting was being held, whom he wanted very much to meet. This man, Gary Karr was considered to be one of the four greatest contre-bassists who had ever lived in the entire history of the instrument. NHM was studying the bass while working in Paris at the Pasteur Institute. He knew of Karr from an article written about him in the New York Times. In his reply to the meeting organizers NHM asked if they could arrange for him to meet Karr. They agreed to this strange request! When NHM was picked up at the airport upon arrival the first thing they said was, we have you scheduled to see Garry Karr tomorrow. NHM marched over to Karr's studio from the Medical School, announced his presence to Karr's secretary and when Karr appeared told him who he was and why he wanted to meet him. It must have been the first time in his career that a scientist came to tell him that he too loved the instrument and hoped to be able to become proficient enough to play it in a symphony. Karr asked NHM about his work in science and how he was able to become a student of J-P Logerot, a highly talented performer and teacher at the National Conservatory, while he was working in Paris. After their first meeting Karr introduced NHM to his students and arranged for NHM to come and play in the Bass studio when he had time between the obligations of the meeting. NHM went each day, and on several evenings. When NHM appeared to play on the evening of the day after he had given his talk the students flocked around to ask all about science. It appears that the local news station had been profiling each day the speakers who were scheduled to give their talks that day and must have made it sound impressive that NHM had traveled all the way from the Pasteur Institute in Paris to participate in the Halifax meeting. In the 30 years that followed, Gary Karr became a good friend and NHM became a symphony player. Gary has even come to visit NHM's lab at the University of Arizona and has seen the beauty of bactertial macrofibers. On his visit to Tucson, Gary related a hilarious story about one of his concerts in Japan. The concert manager waiting for him in the wings attempted to tell him how wonderful his concert had been and how the whole house clapped for him, but mispronounced clapped as crapped. The next day NHM designed the following suit for Karr's next performance in Japan. He called it the crapper suit, so he could keep on smiling and playing even if they did all crap. Gary insisted it be called the Karrapper suit.

OK the Karrapper suit it is. In the picture shown below of Gary wearing the Karrapper suite his face, bow, and the famous Amati Bass he plays were taken from one of his albums. The Karrapper suit itself is actually a hazmat protective suit used for working with dangerous chemicals. Putting it all together is an example of what you can do when you deal with images in your laboratory. Creator who was responsible for the idea initially later was sorry for what he had done. Performer was against it from the start. He argued you do not take a person of Karr's talent and accomplishments and make a joke of him. Nevermind that Karr went along with it. You will be sorry for this he warned. Wait until something like it happens to you NHM he warned. We will see how you like it. Indeed NHM really didn't like it at all when, well wait.

In Paris after the Halifax meeting NHM set about exploring the properties of macrofibers. He rushed to learn as much as possible about their structure, assembly, and growth properties. As word got out about them, people began to filter into his laboratory to get a look at them. Professor NHM explained to them what he hoped to be able to learn from macrofibers and how he planned to go about doing the experiments. It was clear that the plan outlined was far afield from their sphere of expertise but they all seemed impressed with the system. It was interesting to note that each person seemed to see something different in this new system, something that could link their understanding of bacteria to a new multicellular form of bacteria. Costa Anagnastopolus, the man who with John Spizizen discovered DNA transformation in *Bacillus subtilis* and fostered the interest in using this bacterium as a model system of bacterial genetics, came one day to look at the fibers. He was then working in the government laboratory at

Gif-sur-Yvette. Costa marveled at the size and uniformity of the structures. While he was there NHM's host, Dr. J-P. Aubert happened by. Costa looked up at him and said, you better look out, one day you may find them wrapped around his neck like a boa constrictor! People's minds were definitely carried to extremes when they came face to face with macrofibers. This is not difficult to understand if you consider that individual bacterial cells are about four microns long and less than a micron in diameter. Macrofibers reach millimeters in length and two hundred microns in diameter. A single macrofiber can contain a million cells all organized into a single structure.

The microbiologists and geneticists working at Gif circulated their seminar notices to the group at Pasteur where NHM worked. One notice caught his attention. It announced a talk to be given by Professor Benno Muller-Hill, a well-known molecular biologist and scholar working in Germany. NHM hadn't seen Benno since his graduate days in Indiana when Benno was a postdoc in Howard Rickenberg's laboratory. The temptation to pay a surprise visit on him at his talk was overwhelming. NHM arranged his experiments to permit being away the afternoon of Muller-Hill's lecture and travelled to Gif being careful after arriving not to be seen by those he knew so that the secret of his being in the audience would not be ruined. Just before Muller-Hill's talk was scheduled to begin NHM bumped into him face to face in the hallway. Are you Professor Muller-Hill he asked? Yes he replied. He hadn't changed a bit, but NHM must have. Benno had no idea whom he might be. Finally NHM said Benno its me Neil. Neil, it's great you came, he said, may I come to your lab to see the fibers after the talk? Sure, why not? And off they went as soon as his lecture was over. On the Metro back to Paris they filled in details about what each had been doing during the past ten or so years. Benno said he heard about macrofibers from the people at Gif and was hoping he would get to see them before returning to Germany. NHM told him that he had recently found some strains that grew as right handed fibers and other that grew as left handed ones and that he would show him both forms. By the time they got to the lab it was well past dinner time and very few people were around. The ancient building was kind of spooky in the dark. Pasteur himself had worked on chirality of crystals in the same building and when his colleagues at Pasteur heard that NHM had found both chiral forms in bacteria more than one of them pointed out that it must have been fate that the discovery was made at the Pasteur Institute!

In the lab NHM got out his cultures and set some up so Benno could see fiber structure and also determine handedness. Benno marveled at their uniformity and correctly identified them as right-handed forms. Then when NHM went to show him the left-handed structures, they were all right-handed! NHM checked his data book for their history. Indeed they should have been left-handed but when they had been grown under conditions usually used to produce right-handed forms they must have changed their handedness! They looked at each other and both realized that helix hand inversion had just been discovered. You have just found a new dauermodificationen Benno said! What an embarrassing way to make a discovery. You must let me know how this works out he said. I sure will, NHM told him. In the following days and weeks NHM did as much as he could to characterize the inversion process but he really couldn't make much progress on the mechanism involved until he got back to Arizona in the fall and developed the methods to directly observe the inversion process. It is indeed amazing. Fully formed mature fibers of one hand cease twisting in their initial direction when induced to invert and begin twisting in the opposite direction thereby undoing their initial structure. They pass through a state of zero handedness and continue to wrap themselves up in the opposite hand. There is no other known living system that does anything even remotely like this. Pasteur in his tomb beneath what had been his apartment in the neighboring building at the Institute must have been smiling that night when Benno and NHM first saw a new secret of bacterial life.

A note should be added here about Benno's book, "Murderous Science: Elimination by Scientific Selection of Jews, Gypsies, and Others in Germany, 1933-1945", that was published first in German in 1984 then later in English in 1998. Benno must have been one of the youngest people to achieve a high position in academic life in Germany based upon the quality of his research accomplishments. Given his lifestyle and the efforts he went through to avoid having to assume administrative responsibilities so that he could focus entirely on his work, he must have been a difficult pill for his conservative colleagues to swallow. When he came to NHM's lab in Paris eight years before his book on the role German psychiatrists and anthropologists played in the Holocaust appeared, he said nothing of his plans about doing a study of this kind. NHM suspects however that he must have already been thinking about it. The two met in France a second time during the year that NHM was working there, this time by chance at Gif. A colleague of NHM's from the University of Arizona, Dr. David Mount, was in Gif on sabbatical and occasionally NHM and his family spent a day visiting with David and his family. After lunch they usually went for a walk around the beautiful gardens at Gif, which were reminiscent of a painting of the French countryside. On this occasion as they approached the Chateau a figure appeared on a small balcony and shouted, Neil, Neil. They were too far away to see who it was but waved back as they made their way toward him. It was Benno, staring down as if he were the Pope in Rome at his window. I'm staying here for a few days. Can you come up and have tea he asked? Up they went. What brings you back NHM asked him? I am trying to move my retirement to France he explained. It might be possible but the negotiations are difficult. I didn't know you wanted to leave Germany NHM said. Well I have a friend here he said and the discussion moved on to other things. NHM never thought further about that until he heard about Benno's book when it was first published. As far as NHM knows, Benno never did make the move to France. It must have been difficult for him however to stay on in Germany after his book appeared given the state of denial in the country about the role of ordinary people in the Holocaust. Benno's interviews with those who participated in using their scientific and medical skills to foster the atrocities show clearly that they would not accept that they had done anything wrong. His book must have been quite a shock to them and their sympathizers. One wonders, given the resurgence of anti-Semitism now in Europe and particularly in France, whether his planned move would have helped at all to shield him from the hostility his book must have triggered.

Winter took its grip on Paris. On bad days the air itself froze blowing sleet into your face as you tried to walk against the wind. Traditional outdoor markets continued however. In the mail came an invitation to speak at a Gordon conference at Santa Barbara. The group that would be attending would include people NHM knew well. They would be coming from all around the world to participate at this meeting. Usually Gordon conferences are held in New England in the summer but a few are scheduled on the west coast in winter to make travel easier for those working in the west. At the close of each meeting the 100 or so participants are asked to vote on where they would like the next one to be held. NHM supported a group that wanted the California location assuming he would only have to travel from Tucson. In the end he had to come all the way from Paris! But he almost didn't make it. First, he discovered that although his grant funds would support foreign travel there was no provision for a trip that began outside the U.S. and returned to the location abroad. It simply couldn't be done with NIH money. The Gordon conference itself had agreed to pick up all expenses once he arrived but could not afford to cover his travel costs. The University of Arizona wouldn't offer a penny for such use! All other inquiries also came to naught. In his lab at Pasteur NHM was writing a letter declining the invitation when J-P. Aubert, his host at Pasteur appeared. Passing by he stopped in just to say hello and learn of the newest things. NHM told him that he had to turn down the invitation to speak at the Gordon conference because of a bureaucratic snag. That's too bad he said. The next morning an Air France ticket round trip Paris to Los Angeles appeared on his desk courtesy of the Pasteur Institute. The I decline letter then became I'll be there. NHM made plans to stop in Tucson on the way back to Paris to deliver his collection of macrofiber-producing strains to those working in his lab at the University of Arizona.

At the Charles de Gualle airport NHM boarded his polar-route flight from Paris to LA. Almost every seat was filled on the big 747. About an hour into the ten-hour flight a loud noise and whooshing sound came from the side of the plane, the temperature dropped and the plane went into a steep decline. The large side door of the cabin, one row behind his seat and three seats to his left had popped open depressurizing the cabin. Initially the door was pushed inwards. A number of passengers later identified as Air France mechanics rushed to the door and held it from pivoting to the full open position. They tied it in place with coat hangers and blankets. The captain came on the loudspeakers and explained that they would return at low altitude to Paris. After landing no one was let off the airplane. Some mechanics boarded, came to the door and examined it using a flashlight and a screwdriver! They opened and closed the door several times and left. The captain announced that they would take off again. NHM couldn't believe that nothing had been done to examine the locking mechanism. This time about thirty minutes out of Paris the captain announced that the cabin was not pressurizing normally and so they were going back to Paris again. The stewardesses announced a Champagne party. When they got to his seat NHM refused the wine and asked for his ticket. At the airport he went to TWA and told them he wanted to switch his flight because of what happened. They gladly accepted the Air France ticket and scheduled him on the next flight six hours later. On TWA NHM noticed that all doors had an electronic device attached to them and were tested prior to takeoff. Nothing like that was on the Air France aircraft although they were both 747's. The TWA flight arrived in LA sometime the next morning. NHM was feeling very ill. He claimed his bags from the original Air France flight and walked out into the airport.

On the horizon beyond the parking lots NHM could see a large hotel. He dragged everything he had with him to it and told them that he had to sleep immediately. They led him to a room where he collapsed and didn't move until the next morning. When he awoke he found himself in a luxury suite, marble bath, the works. Get me out of here, he thought, my whole grant will be spent on this. He checked out as quickly as he could, got something to eat and decided to take a bus to Santa Barbara. It arrived there mid morning. NHM tried to find a cab to the resort at which the meetings were to be held but couldn't. He phoned the resort. It was New Years Day they informed him, guests for the meeting weren't expected until the next day. Their courtesy pickup driver had the day off so no one could come and get him. He finally saw a cab and forced him to stop. In it was the driver with his family going to visit relatives. NHM told him he must find a cab. The driver said he would try to call someone. Two hours later a cab arrived. At the resort a few people who had come to the meeting from Japan and Australia were already there trying to get over their jet lag. The meeting itself turned out to be a dud. LA to Tucson went fine although the visiting professor's quarters that had been reserved for NHM at the University of Arizona turned out to be in a locked building that he could not enter. From Tucson it was New York, then Paris. He was so happy to be back that he hardly noticed that the weather was miserable. Later Joan told him it had been much worse during the time he was away. Let's hope for an early spring, he said.

As summer finally approached those working with him at Pasteur told NHM that the city became very unpleasant as the temperature rose. Paris would be congested with tourists and could even at times begin to stink they said. Everyone who could left to go elsewhere for as long as they could afford to stay away. Indeed signs soon began to appear in shops announcing the dates of closure for annual vacations. Some places would close for a week, others a month. Even the scientists at the lab were busy making arrangements to get away. NHM decided that perhaps he too should leave for the summer. His schedule required that he return to Arizona in early fall but nothing required that he return from Paris. Why not move to another lab and work for a few months in a better environment for his family? NHM telephoned people he knew who might be in a position to act as hosts at other institutions and accepted a very gracious offer from his former colleague at the MGRU in London, Dimitri Karamata. Dimitri had returned to Switzerland and become the head of an institute at the University of Lausanne. Not only did he offer space, but also he volunteered to find a flat that they could live in during their visit. Lausanne was only a day's drive from Paris. Summers there were cool and the setting on Lake Geneva was idyllic. Dimitri's Institute was housed in an old building much like the one NHM had been working in at the Pasteur Institute. The space was luxurious and well equipped. NHM made arrangements with his technician in Paris to finish some experiments before she took her vacation, and to then send his cultures to Lausanne, and another set of them to Arizona. NHM rented a station wagon into which he crammed all of their goods including a double bass that he had purchased from his teacher in Paris, a student cello bought for one of his daughters, and a viola that his other daughter had taken with her to Paris. In addition to all that a fifth person came along with them, a friend of the children's from Tucson who eventually moved to Europe for her adult life. In spite of it all, the trip went smoothly. NHM called Dimitri when he reached Geneva and Dimitri met them at the apartment building where their flat was located. It turned out to be in the same building where Dimitri lived. A Swedish woman who went to visit her family in Sweden each summer had agreed to rent her apartment to them. It was a beautifully furnished place just north of Lausanne in a village called Epalinges.

Epalinges is now a village of about seven thousand five hundred people. When they lived there it was still rural enough to have open space and at least one active farm. As the village flag shown below suggests it was an uncomplicated place, orderly, manicured, and secure in its peaceful existence.

Then the Americans arrived. Not too many visitors made their way to Epalinges so they were immediately noticed, and carefully scrutinized. Within a week there came a knock upon the door. It was the local police. Very friendly, very formal, they asked if NHM knew anything about a parked car in the neighborhood that they were unfamiliar with. No, we have no car NHM told them. Thank you very much sir, they said and away they went. The next day Dimitri told NHM that the car in question had been abandoned in Epalinges for months before their arrival. A similar routine occurred every time they came for the summer to Epalinges. Each time it was something different, all insignificant and simply the Swiss way of informing foreigners that their presence was known. Once NHM got established in the lab, which was a fifteen-minute walk from his place, or five minutes on an electric bus, he unpacked his bass and began to play for an hour after dinner. He had some excellent Bach and Mozart transcriptions that he was working on. The third evening of playing ended with a visit from Dimitri. Look, he said, one of my neighbors has come and said that her husband cannot enjoy his dinner when you are playing the bass. Please change the time you play, he said. Should I play later, NHM asked him? Well probably not he replied, the rule is no noise after 8:00 p.m. All right, I will find another time, NHM said. The next day NHM went home after lunch and played through a Mozart piece before going back to the lab. Again Dimitri appeared after dinner. Please he said, this woman's husband takes a nap after lunch. Don't forget that after you return to Arizona, I will have to still live here. O.K. there would be no more playing during the week for NHM. What about on the weekends? I'll find out Dimitri said. But the truth is that they had so many things to see and do on weekends that NHM couldn't find any time at all to play. Dimitri was very happy. NHM did point out to him that the gander's mouth on the village flag was open as if it were singing, and that perhaps granted a right for some music. Perhaps, he said, but the villagers would prefer that it be confined to the Music Conservatory in Lausanne or the Music Hall. The American invaders had finally understood the culture in Epalinges. Silence reigned.

NHM established macrofiber cultures at the Institute in Lausanne and pursued experimental and theoretical work on their structure. He was particularly interested in what their interior organization was like assuming that a clue to their morphogenesis might be found by knowing the geometrical arrangement of the cell filaments on the inside as opposed to the outside of the fibers. Measurements NHM had taken of fibers in Paris suggested how the interior might be organized but no one had yet actually seen the inside. Dimitri, originally an expert on electron microscopy, made some efforts at visualizing the interior but fixation problems and other technical difficulties prevented him from getting the resolution needed for model building. Meanwhile NHM worked out the relationship between fiber diameter and the pitch angle of filaments lying on the fiber surface. From that he concluded that filaments at the center of a cross-section must lie nearly parallel to the long axis of the fiber while those on the surface would be nearly perpendicular to it. NHM was unaware at the time that this organization is well known to textile fiber experts. Multifilament twisted yarns obey the same rules of geometry as do macrofibers!

On weekends Dimitri was anxious to show them the wonders of Switzerland. They met many people and went to the tops of mountains, places they should have never visited given their lack of mountaineering skills. At lower altitudes they took a short boat ride across Lake Geneva from Lausanne to the French town of Evian, the place where bottled water is produced. Presumably because of the water quality the town is a center for kidney specialists with many clinics that serve patients from all over Europe. In the center of the town there is a spout coming out of the side of the mountain that has been continuously running for hundreds of years. It is fed from the glaciers and snowfields atop the mountain, which must also be the source for the bottling company and the nearby health resorts. NHM decided that on their next visit they would bring some plastic cups with them so they could all taste the water. When there some weeks later a short queue of people were waiting their turn at the spout. Immediately behind them on the line was a couple from Japan. The husband had an enormous video camera with him and was filming everything in sight. They were obviously tourists having the time of their lives. Finally it came our turn to draw water. NHM went up first to sample the famous fluid. He bent over to fill his cup, straightened up then drank its contents very slowly. As soon as the water went down he began to puff himself up as if the water had transformed him into some kind of hulk. He raised his hands above his head and turned to face the camera directly. At that point Joan and NHM's children ran away laughing. The Japanese man kept filming without stop. The look on his face should have been captured on film. He and his wife were totally perplexed. Shocked would be the right word. They knew NHM was not a local person. His cowboy hat didn't match anything others had on. NHM stood in place for several minutes.

Everyone on the queue stopped talking and watched carefully for what he might do next. Then he began to wander off as if in a cloud. No one said a word. He never looked back but his daughters later told him that the Japanese woman eventually went up to drink her sample, while her husband took the pictures. On the way back to Epalinges they discussed how the Japanese tourists might explain this incident when they were back in Japan and he was showing his movie of the world. Why did you do that NHM's children asked? He wished he knew the answer but he didn't.

7. How are you going to keep them at the U of A after they've seen Paris?

As fall approached NHM made flight arrangements from Geneva to New York. He began to think about how to get the musical instruments back with them. Dimitri suggested that he ask the father-in-law of one of his students, Barblon, who was a cabinet-maker, if he could build a box to transport them in. Barblon, volunteered to take NHM to the shop and to discuss the matter with his father-in-law. When they arrived at his shop the old man was anxious to be helpful. Together with NHM they designed a pine box in which all three instruments could be safely packed, with carrying handles at the ends and a removable lid. He assured NHM that the box would be very lightweight and sturdy. A week or so later Barblon came to NHM's lab to let him know that the box was almost finished. He said that his wife's father wanted to make sure the dimensions were correct before finalizing it. Could professor bring the bass to his shop for a fitting and help decide about the packing insulation and lid design, he asked? The next day they toted it over, laid it out in the box and decided that the lid would have to be attached to the walls with a series of screws that could be removed at customs in New York. The cabinet-maker said he could complete everything that day and volunteered to deliver it to NHM's apartment building along with the bass on the following day. It was a done deal. NHM went to purchase a screwdriver of the appropriate size so that he would be able to take the bass out and up to their apartment until the evening before their departure, then reattach the lid when everything was secure inside. His plan was to carry the screwdriver in his briefcase on the flight in case he had to open the box upon arrival. The box with bass inside arrived at their apartment-house mid-afternoon the following day. NHM was surprised at how light it was when he and the driver carried it into the lobby. They set it down in the center of the lobby and NHM went to get his wife to help him undo the lid and remove the bass.

The new wooden box looked exactly like a coffin. People coming and going were quite apprehensive about it. The concierge of the building came by and asked if someone had died. No they hadn't NHM told him and he went on his way. The neighborhood children discovered the coffin-like box and saw that it's lid was about to be removed. They quickly assembled all their friends. A crowd of thirty or so surrounded Joan and NHM. One child's mother appeared as they neared the end of their task. Each screw that NHM removed was counted and handed one by one to Joan who put them into a plastic bag. There were eighteen screws in all. NHM said, only three more. All the children stared intensely at the lid. The mother watching it all became terribly anxious. She grabbed her son's arm and began to drag him out of the building but he wouldn't go. She had to let go of the child but the suspense was too much for her. She ran toward the door afraid to look at what might be in the box, but also afraid to abandon her son. Finally screw eighteen came out and NHM passed it to Joan. The place became dead silent. NHM stood up, grasped the edge of the lid, paused to look at the children, then lifted it only a tiny bit as if to peer in. He immediately slammed it back down as if horror was in there and he shouted loudly, Dracula and Frankenstein! The woman ran out of the building overtaken with fright. The children went wild, jumping and shouting as if they had indeed seen the two. NHM took off the lid. The children pushed up closer to look in. At first they could not see exactly what it was because of the insulation packed around it so they all became silent again. When NHM unpeeled the bass the children had the biggest let-down of their young lives. Out they went to play while NHM and Joan took the bass up to their flat and stowed the coffin behind the staircase in the lobby. Neither of them ever saw the poor woman who ran away in fright after that incident although NHM believed she was the woman whose husband had objected to his playing the bass altogether. Who would have thought that just putting the bass in a box could be worse than having to listen to it?

When departure day arrived a truck pulled up in front of their apartment building. It belonged to the Beckman Instrument Company in Geneva, a well-known American company from which NHM had purchased several large centrifuges. During his stay in Lausanne NHM met the head of the Geneva office of the Company. When he learned that NHM was trying to find a way to get his large box to the airport at Geneva he offered to have one of his delivery trucks take it there for him. Together with the driver NHM loaded the packed coffin-like box onto the Beckman truck. He then loaded their suitcases into Dimitri's car and Dimitri drove the four of them plus his wife, Jasna, and all the baggage to the TWA passenger loading area. NHM transferred the box from the Beckman truck to an airport baggage cart. Of course the box was too long to lie on the cart as a suitcase would, so he stood it on end. It looked odd to see what appeared to be a coffin proped up on its end. Other passengers appeared perplexed at the way he handled it. The TWA people, however, knew it was coming. The head of the TWA Geneva office was also a friend of Dimitri's. He had agreed to airfreight the box on the same flight they were taking for a fee equivalent to that for three extra suitcases. It was all arranged beforehand. The TWA baggage handlers took the box from them at the check-in counter and rolled it out a door that was normally used for freight, just as if it were a coffin. Everything went smoothly. When finished with the check-in process they thanked Dimitri for his hospitality and got settled in for a three-hour wait until their flight would depart. Dimitri asked, why wait here? Lets go into Geneva he said, I want to show you something. So they all climbed back into his car and he drove them to a lakeside park. There they strolled toward a chateau situated right on the edge of Lake Geneva or Lac Leman as it is known locally. When they reached it Dimitri said, let's go in. NHM saw that it was a science museum and that it was closed. Dimitri paid no attention to that. He walked up to the door as if he owned the place. There a woman, the curator of the museum, greeted him. She was a scientific colleague of Dimitri's and had agreed to open the museum for our visit. In they all went for a private tour. Although it was a small collection there were many wonderful things to see. They soon came to one of the first electron microscopes built in Switzerland that had been used for biological purposes. On the control console of the instrument there was a small metal frame that held a hand written card giving magnification ranges for various settings of the dials. NHM recognized it immediately as Dimitri's handwriting. Before Dimitri came to London to do his Ph.D. he had worked on the construction and use of this instrument under Professor Kellenberger's direction. NHM called everyone over to the instrument and asked if they recognized anything familiar on this machine. No one did. Then he pointed out that Dimitri had written the instructions for its use. Yes I did, he admitted. Immediately his wife began to kiss him. She herself had no idea that something of his was in the museum. Probably few people would have recognized his handwriting but NHM had worked closely with him in the same lab in London and knew it well. Back at the airport as they boarded NHM's children said that was very impressive. It was indeed.

The flight back to New York was uneventful. Fortunately they all were well rested when they arrived for the events that transpired at Kennedy Airport were anything but uneventful. After passing through passport control they proceeded on to baggage claim and customs. All their bags had arrived but not the coffin. They waited until all the other passengers from their flight had cleared customs then asked where the coffin was. It is upstairs one floor above us they told NHM, where everything is taken from the planes before being transported to the customs floor. Why hasn't it been brought down with the rest of our luggage? It will not fit into the elevator, they replied. How then do you get coffins down here he asked them. Coffins do not come as baggage they said. They are sent as freight. Freight passes through another building, not this one. Well why don't we go up to the box and inspect it upstairs, NHM asked. The law is that the airlines must present the baggage to US Customs; Customs does not go to the baggage they told him. Well how does the airline expect to do that in this case NHM asked. Of course they did not know. Neither did the airline. Why don't you carry it down the stairs, he asked, it is very light. We tried they said. It did not fit going around the turn in the stairwell. In the meantime baggage handlers at Kennedy became aware of a problem with a coffin. They read on the shipping label that the coffin belonged to Dr. Mendelson from the University of Arizona Department of Microbiology and Medical Technology and assumed that he must have tried to bring a corpse from Europe with him for some kind of weird experiments. Several of them asked NHM about it. He told them there was no body in the box, and there wasn't anything to fear. At least one of them told NHM that he wasn't going to carry it even if it got through Customs. In fact he wasn't even going to touch the box. Another asked NHM why it didn't come in that special sealed container that bodies are usually sent in.

They waited for another hour. There was no resolution. It came time for the next shift of the Customs Inspectors to assume responsibility. NHM explained his situation to them. They sent him back to the airline. Finally he made a suggestion to them. If they would let him go to the box on the floor above he would open it and bring the contents down to the inspectors. They had no guidelines for this and finally decided that if an inspector came to watch him unload the box, it might be all right. An inspector was assigned the task, and with him a baggage handler. NHM took the screwdriver from his briefcase and the three of them went upstairs. There stood the coffin in perfect condition just as he had left it in Geneva. The baggage handler became very nervous but the Custom's official was anxious to see the contents. He said the last shift workers had told him all about it. NHM went through the process of removing one screw at a time just as he did in the apartment house in Epalinges. Finally off came the lid. The inspector started to laugh, but the baggage handler refused to look. Go ahead the Inspector told him; it's not a body. The poor man peered in then shouted, "It's a bass man, it's a bass" and off he ran never to be seen again. The inspector said don't bother to bring the contents down, I will clear it for you up here. Drive your car up to the freight pickup to claim it. Once their coffin was neatly tucked into the station wagon they had rented they transported it to NHM's parent's home in the New York area. It was a very long day but everything was not yet over. Arrangements still had to be made for the shipping of the coffin back to Tucson and their own flights home as well. Nothing could be done however until jet-lag ran its course and energy metabolism was restored. Even so, Performer was not silenced for long. He insisted on an abrupt termination of the French/Swiss lifestyle and its replacement by what he called “the American Way”. No soft landing here.

The long flight from New York to Arizona was really too short for the mental tasks required to bridge between what had gone on in the year now ending and what it would take to piece together the life left behind in Tucson that would now have to be resumed. There was very little said to one another during this flight. Each of them had a separate job to do in trying to knit together what it had been like before, where they were now in their own development, and how they hoped to move forward. Two teenagers and two adults were no longer what they had been and they knew it. No it wouldn’t be like coming home after a short break. Nor would it be like starting again. None of them could really see clearly however what it would be. That was the frightening part of this return journey. In NHM’s mind Creator kept asking Observer for details about the events that had gone on in the days and months before departure for Paris. Neither he nor any of the four NHM components had given any thought to these things for the past year. Recent events had taken over their focus and now it was time to reawaken memories of older things that would be needed once on the ground in Tucson. Observer did his best to dredge up all the information. And so the dialogue spun within NHM’s head probably at the rate the jet engines spun the air to move the airplane through time and space.

When NHM finally got back to the University of Arizona from Paris, much to his chagrin, he learned that the Dean had not acted upon the report that had been presented to him a year earlier concerning the review of the Microbiology and Medical Technology Department. Nothing had changed. The same department head was in place and NHM would have to return to work under his authority after having penned the report that concluded that he should be removed. NHM went to see the Provost about this and explained to him that he did not want to remain in the MMT Department. He did not want his salary recommendations to come from the head of MMT, and he did not want to work in the MMT program. Where would you like to work, the provost asked? Could I perhaps work in the newly organized Arizona Research Laboratories, NHM responded? This was a newly created administrative entity about which little information was circulated. NHM had the impression it was meant to foster research and development. It would be better the provost said if you could work in a department. Would you consider moving into the department of Cellular and Developmental Biology (CDB) he asked? We created this department to strengthen research in life sciences he went on. NHM said he would look into it and get back to him before making a decision because he knew very little about that new department. The CDB department head, Dr. H. Vas Aposhian gave NHM a brief synopsis of his department, its origin, objectives and privileges. He said the administration had brought him to Arizona to build a strong research group in modern biology. It would be a place where distinguished younger investigators would be sheltered from older faculty who had very little research accomplishment and were not sympathetic to either the research culture, or the rapid advancement of those with research productivity. In this way older faculty who had been hired at Arizona in an earlier era and for different purposes would not hinder the new stars. There were some older professors in CDB but the department's focus definitely was on research productivity. Very little teaching would be required of CDB faculty.

Dr. Aposhian was aware of NHM's research program and told him he would be very happy if NHM wished to transfer into CDB. NHM knew one of the young professors in CDB, Dr. Marty Hewlett. Marty had been a graduate student in biochemistry at Arizona and NHM had him as a student in his Advanced Microbial Genetics class. After his Ph.D., Marty went to MIT to do a postdoctoral with Dr. David Baltimore, a highly respected Nobel Prize winning molecular biologist. Hewlett called NHM just before he left for Paris to tell him that he was going to return to Tucson and asked if NHM was going to rent his house during the year he was away. He wanted to find a place in a neighborhood where he, a black man, and his white wife and their son could live without being hassled. NHM thought the neighborhood where his home was located would be ideal and it proved to be so. Marty, too, thought it would be a good idea for NHM to move into CDB. NHM trusted his judgment and informed the Provost that a move to CDB would be acceptable to him. No physical move was involved only an administrative one. As a professor of CDB, NHM found a warm reception within the group. The older professors knew that he had a classical biology background and that he could talk with them about their interests, whereas the younger people viewed him as a jet-set modern biologist working at the best possible places and on topics in which he was the leader. So it all went along very well until one day unexpectedly Vas Aposhian announced that he was submitting his resignation because of a disagreement with the central Administration.

It was late in the spring when that happened, at the end of the academic year. NHM had arranged to go back to Switzerland for the summer to work on a project that involved new strains of *Bacillus subtilis* that others found also produced bacterial macrofibers similar to those he had produced in Paris. A phone call interrupted NHM's thought processes then focused on a new kind of cell timing model that he had developed. That wasn't the only thing this call interrupted however. It led to almost five years of major distraction and difficulty. The secretary to the Dean of the College of Liberal Arts, the very same man to whom NHM had submitted his review of the MMT department was on the phone. She said that the dean would like to meet with him and asked if she could schedule an appointment. NHM provided her with the times he had free and she selected one that fit into the dean's calendar. There was no mention of what he wanted. NHM thought it would either be about the MMT Department report, or possibly the establishment of a search committee to find a replacement for Dr. Aposhian. He was wrong on both counts. What he wanted was to know if NHM would be willing to serve as acting head of CDB until the eventual fate of the department could be determined by the central administration. NHM told him that he was about to depart for Switzerland and thought it impossible but to his surprise the dean said. "That shouldn't matter." How could that be, NHM asked? Over the summer there should be very little that you would have to do, nothing that you could not do over the phone, he said. "Are you sure", NHM asked him? Yes indeed, he replied. How would the others in the department feel about it, NHM inquired? "They like you", he said; "they would be comfortable with you as acting head". "Would it be all right if I let you know tomorrow", NHM asked? Yes it would he said. NHM and Joan discussed it at home that evening and agreed that perhaps NHM should help out CDB. They were kind enough to provide a shelter for NHM a year earlier, and so NHM notified the dean that he would assume the responsibilities and it was arranged. Of course, there were some major disputes in CDB that had to be resolved from Switzerland over the summer but it wasn't really difficult to get them settled. Most of NHM's attention while at Lausanne was directed to his discovery that one of the new macrofiber-producing strains contained in the Lausanne collection underwent a helix hand inversion as a function of temperature. It was a strain isolated by Dr. J. Fein when he was working as a postdoc with Howard Rogers in London, and given to Dimitri before he moved from London to Lausanne. NHM determined that the transition temperature was 39 degrees C: decreasing temperatures at which structures grew below 39 caused the production of tighter right-handed structures, whereas increasing the growth temperatures above 39 led to tighter left-handed structures. At 39 degrees structures formed with zero twist. This temperature system provided an easy way to study the regulation of twist and handedness in macrofibers. That was the main focus of NHM's efforts for the following several years.

At summer's end back they went to Tucson from Lausanne like yo-yos. Once there, things started to heat up. No, it wasn't just the air temperature. The dean called NHM to let him how well things went in the CDB department during the summer. He asked him to consider continuing as the acting head of CDB until a final decision was made about how to resolve the leadership vacancy. How long will that be, NHM asked him. Not too long he replied perhaps a month or two. Yes I can do that NHM told him but there will be no change in the way I will run things. That will be fine, the dean said, and thanks. So NHM moved some things into the department head's office in case he had to meet with anyone concerning department business, but kept everything intact in his true office located with his labs in the Shantz Building. NHM held a department meeting to give the faculty a chance to let him know about what was on their minds. Several professors were concerned about whether CDB could recruit new faculty according to the original department plan. NHM said he would look into it and told the faculty how he planned to operate as acting head. First, he said, I will not have any unnecessary meetings. If you have something you need to know go first to the administrative assistant and see if she can solve your problem. I will be in my lab, office, or giving lectures all day until 5:00 p.m. From 5:00 to 6: 30 p.m. I will be in the department office doing administrative things. If you need to see me please schedule an appointment for after 5:00 p.m. Any questions, he asked. No one had any. NHM chose to operate in this way for several reasons. Foremost was his need to live life as a research scientist not an administrator. NHM had no desire to manage anyone or any program for that matter. All he yearned for was the opportunity to go off in his corner and quietly make discoveries about how living things did their things. Guardian knew this and fought against diversions.

The first person to make an appointment was Vas Aposhian, the former head of CDB. He said that the administration had made a commitment to him and the CDB department to hire several new faculty members and that NHM should make sure that they did not take back those budget lines. O.K., NHM said, I will call the dean about this. When he did, the dean said he would find out from the provost and get back to him. A week later the dean gave the go-ahead. The provost told him that NHM should definitely try to hire a new person that year and perhaps two the following year. That sounds good, NHM said. We will start on it soon. One other thing the dean added, the provost suggested that we should make you the permanent head. I am ready to do so, how do you feel about it? My goal in science and as a professor was not to become an administrator, NHM replied. You know I do a minimum as acting head. I can't give up my scholarship to run a program much as I would like to help our department and the university. You won't have to give anything up he said. We think the job you are doing is excellent and would be very happy if you just continued in the exact same way. Do you really mean that, NHM asked him? Yes I do he responded. Think it over and get back to me this week. What else could NHM do? He called a faculty meeting, told them what had happened and outlined the conditions that would have to be met if he were to accept the offer. He told them that there would be no basic changes from the way he had worked as an acting head. They knew his priorities. He asked if anyone else would consider becoming the department head. One older professor said he would consider it. NHM told him he would inform the dean of that and find out his response. When he told this to the dean the dean suggested that it be put to a vote of the faculty to determine which of the two they would prefer as their administrator. Then he said I want you to win that vote. NHM knew he was in trouble then. He definitely was not going to lobby for votes. He knew that if he lost there would be no place in the institution that he would be able to hide from the consequences. It would be as if he had given everyone a vote of no confidence. Finally NHM decided that the only thing he could do was to vote for himself rather than to vote for the other candidate or to abstain. In the end NHM won by one vote. Although he felt kind of sick about it he tried to talk himself into the idea that by being the Head of the department he would be able to protect himself against things better than he could as simply a Professor. In the end NHM's tenure as head of CDB lasted about four years. He resigned when a new university President arrived who had a very different view of how the biological sciences should be organized than he did. Resigning was probably a stupid thing to have done but Guardian was as pleased as he could be. Creator and Performer celebrated. Observer tried to compress all of the information he held about NHM's tenure as department head into the smallest possible space. Only now has some of it been decompressed to fill in the story.

NHM must have been the worst department head the University of Arizona, perhaps any university for that matter, had to endure. He simply could not swallow the unethical and corrupt things that went on in the running of the organization. Dishonesty abounded. People were suckered into doing things simply to give the appearance that there were inputs into decision-making that never existed. Arizona was anything but an Academy of Scholars seeking to find truth and to communicate it. It was a place for political agendas to promote favored sons with position and money. Team play was the only thing. Thinking something through was definitely not desirable. Nor were new ideas. To protect the CDB department and himself as well NHM did the following. He hired a new Administrative Assistant and gave her the responsibility of dealing with all trivial matters. He left it up to her to determine what was and was not trivial. This worked well from NHM's perspective but made others very angry. One day an associate dean called and insisted that he be put through to speak to NHM. She explained that NHM could not be interrupted when he was working in his laboratory but she would be glad to help him. He was reticent to say what the call was about. Finally she told him that she had responsibility for dealing with all trivial matters. He hung up before she could say that if he wouldn't tell her what he wished to know or say, she had no way to judge whether or not it was trivial. At a later date he called again to speak with her. This time, he told her that he wanted her not to do as NHM had instructed about a particular issue but rather to do the opposite as the dean's office wished. When she told NHM that later in the day he called the associate dean's boss, the dean, and told him to make sure nothing like that ever happened again, and that he would have nothing further to do with the associate dean. The dean apologized for the misbehavior of his associate, and said it would not happen again. Once NHM's management style became known, people tried to find ways to breach it. In the most flagrant cases NHM would call directly to the offending party to say that he did not want to hear anything again from them about the matter. I know your position and do not value it he would tell them. That often worked. The second thing that he did during his tenure as head was to accept invitations to serve on high-level central administration committees where programmatic and higher level decisions had to be made. These were the most revealing opportunities that he had to see what it is really like running institutions of higher learning. NHM did everything he could on those committees to fight against corruption but it came to naught. People's careers were ruined by the machinations that took place at those meetings. People's careers flourished by the machinations that took place at those meetings. Whether you got the former or the latter treatment had nothing to do with your abilities, accomplishments, or experience. Power and connections were all that mattered. Those meetings were the most difficult things the Guardian ever had to swallow. They made him sick. They made him yearn for the solitude of the research laboratory. Creator told him, oh, grow up. That is the way the world is. The other partners agreed but still all four of them couldn't wait for it to be over.

At the time NHM first became affiliated with the CDB department it was part of the college of liberal arts. When he became its head he was required to attend periodic meetings at which the dean discussed college-wide issues with all the department heads. These meetings were just about the only times in his thirty-five year tenure at Arizona when he had an opportunity to meet and speak to people who weren't scientists. To his disappointment he found that the majority of the other department heads had little or no interest in what was going on in other disciplines. Their focus was only on whether they were getting their fair share of the pie: money, space, and every other perk known to man. Most of them were fairly hostile towards the sciences and their practitioners in part because they viewed them as arrogant and financially privileged. Put simply they resented the fact that the pool of outside resources available to science was so much greater than it was to them and that the central administration valued money above all else. The meetings were indeed dour, and the fact that the dean had mainly bad news concerning the funds that the State awarded each year to the University didn't help matters. NHM tried to interject some humor from time to time into the discussion but it was clearly out of place. The dean, himself a linguist/historian/Latin-American scholar was highly promoted by the university president as the ideal "Renaissance man" to lead the college, the largest administrative entity of the university. He was viewed as the heir apparent to the provost's position. The department heads had different opinions of him however. Those who were not in the sciences took pride in the fact that one of their men was in charge of us, whereas the scientists were frequently frustrated by his poor understanding of their culture and the kinds of competition they had to face in order to survive.

Eventually the president of the university, Dr. John Schaefer, himself a chemist, came to the conclusion that the transition to a research oriented university would necessitate the reorganization of the colleges. He felt that the sciences would have to become their own fiefdom, if for no other reason than to gain total control of the funds they brought to the university. He never actually said this in public however. To set his plan into action the president created a committee charged with developing a plan for the college of science. NHM was asked to serve on it.

A man the central administration hoped would become the first dean of the college of science chaired the committee. As NHM got to know him he couldn't understand why this man was their pick. He appeared to be a man with virtually no scientific achievements, and a rather shallow grasp of the complexities of the scientific world. NHM thought the chairman was totally unsuited for the job but listened to everything he had to say and tried to correct the things he got wrong when final recommendations had to be written by the committee. The plan that was produced after a year's work together with whatever it was that the central administration put on the table resulted in approval at the State level for the proposed reorganization of the University of Arizona. A search committee to find the first dean was quickly appointed. Yes, NHM was one of the anointed that served on it. The head of the committee was, as you might expect, a chemist given the fact that university president had also been a chemist. Would you be surprised to learn that the chairman of the committee that planned the college submitted his application for the deanship? So did a lot of other people both from the UA and other institutions. Some of them had very distinguished records of accomplishment and they naturally rose to the top of the list as the committee members ordered the more favorable candidates from those less likely to be acceptable. The man the central administration favored did not make the cut into the top twenty five applicants. Before the committee could make its final recommendations however the president of the university announced that he was going to resign as soon as his successor could be found. It appeared to NHM and others on the dean's search committee that the proposed university reorganization was too far along to be reversed even if the central administration was going to be changed. The committee continued winnowing down the applicant pool until they had their top three to five choices selected. The central administration contacted each of them. None of them was willing to come to the University of Arizona even for an interview. The imminent change of the central administration was a factor none of them would risk. How could anyone move to an institution to fill a dean's position without knowing who those above him would be? Creator knew this. His fear was that all the work done to find a qualified candidate might be lost and that as a matter of expediency the position might go by default to an undistinguished person.

During the lame duck period of Dr. John Schaefer's administration his operatives continued to function in their old style. NHM got a call from the vice-president for research. The caller asked him to please block an initiative on which the dean of the college of liberal arts was moving ahead. He told NHM that the central administration did not believe it would be beneficial to the new college of science. "You know" he said, "Paul doesn't know anything about science and he is way off base on this". When NHM looked into the matter he could see no reason for the vice president's request other than that it would cost something. The central administration must have been searching for money that they could divert from other programs to the new college of science. NHM was outraged that they should have asked him to do their dirty work. Had they already forgotten that they had recently cheated NHM out of a large amount of money? NHM had been awarded a so-called, "decision package" by the State of Arizona to start a new program in structural biology within the CDB department. In a competition that began at the program level then advanced through the system NHM's application was one of two or three that were forwarded by the University of Arizona to the Board of Regents. It received top priority by the Board and was funded by the State Legislature, but the money never materialized. When NHM sought to learn where the money was, he discovered that it had been appropriated by the central administration for use in establishing a totally different program, a Neurosciences program that would be housed in the Arizona Research Laboratory. Guardian screamed bloody murder at the theft. He insisted that NHM should force an investigation. But he wouldn't. Why he didn't blow the whistle on that is a complicated matter. The fact that he didn't has plagued him forever. What better insult could there be than for the central administration to ask him to do the same thing in turn to someone else? NHM told the vice president for research that he would look into it but did nothing. It didn't matter. The college of science was formed and the man they favored did become the founding dean as unbelievable as that sounds. Creator had gotten it right. It wasn't a smooth transition however. The vice president for research who attempted to enlist NHM's support flew the coop to another institution never to be heard of again. Think of what it must be like at the institutions that recycle these less than outstanding administrators from other universities. But wait, could the same happen at the University of Arizona?

The new University of Arizona president was named. It was Dr. Henry Koffler, a former University of Arizona student. There was much local press about this along the lines of: one of our own returns as president, a homecoming for Dr. Koffler, and so forth. NHM's phone rang off the hook. Congratulations were the message of the day from colleagues at the University of Arizona. They seemed to assume that Koffler's coming home would benefit NHM or microbiology or both. How strange Observer thought. NHM knew Koffler only superficially from some of his research publications. The two never crossed paths either at prestigious research institutions where NHM worked or at meetings that NHM attended to which leading contributors were invited. Koffler's publicized move to Arizona also brought a wave of inquiries to NHM from scientists around the country who sought information about the availability of jobs at the University of Arizona. Many of these individuals held administrative positions elsewhere but for some reason the grass looked greener to them in Tucson, in spite of the desert sand, than it did wherever they were living. Why do you suppose they would want to come here, Observer thought? Certainly it couldn't be Arizona's reputation as a center for microbiology, or related life sciences Creator chimed in. No, it must have been something else. Perhaps they want to ride in on Koffler's tail because they share with him a scientific culture and think they would fit well into the kind of programs Koffler would foster. That could be it NHM thought but the theory could never be tested for the initial adrenalin rush to get here was never followed by the arrival of any of Koffler's admirers. For one of them however the attraction to Arizona proved a fatal attraction for his career in research. It is a sad story but worth sharing.

When NHM received the initial call from this individual he could have never guessed there would be anything unusual about it. The caller was a person NHM knew superficially, had met long ago and remembered as a friendly pleasant person. In many ways he was no different from all the others who wanted to know what it was like here and whether there were any job openings presently or expected after Koffler's arrival. After listening to NHM's reply the caller thanked him for the information and said he would call back if he wished to pursue anything further. That was pretty much what all the callers said. But unlike the others in this case the caller did call back, again and again. Each time he wanted some further details. Although he never gave any details concerning the kind of position he sought, NHM got the impression that he was after a high level job in the central administration. Creator came to the conclusion that this individual must have been a close friend of Koffler's. Could it be that Koffler was going to create a special position for him, NHM wondered? Guardian counseled that if it were so, NHM might eventually come under his control at Arizona. Fortunately that did not come to pass but it might have.

NHM thought for sure that this caller was headed to the University of Arizona when he learned that he had been invited to come to Tucson for an interview. I would like to come by to see you when I am there he told NHM. Then he asked a favor. Neil, he said, could you send me some information about your most recent research? I am sure you will come up in my conversations with the central administration and I would like to tell them about your great current findings rather than things from olden days. I would be happy to send you recent papers, which would you like, NHM asked? Whatever you think appropriate he responded. I have just finished a review that is now in press NHM told him, would you like a preprint copy of it? That would be ideal, he said. This paper should be out soon, NHM went on, if anyone asks you where it was published tell them I will send them a copy as soon as it comes out. That sounds perfect the caller added. And thank you very much for everything. You are welcome, NHM continued. I will mail the preprint to you tomorrow. I am looking forward to seeing you soon in Tucson, was the way the discussion ended. Not long thereafter Dr. X did indeed arrive at NHM's office, told him how much he enjoyed his visit and how highly the administration thought about NHM's work. Then he vanished into the sky flying back to his home with never a mention of whether he was moving to Arizona and if so into what job.

There it remained for months, and possibly forever had it not been for what arrived in the mail. It was a research grant application that had been submitted by this same individual to a national granting agency and then sent to NHM for review. Although it did not deal directly in any way with the kind of things that NHM studied it was nevertheless the kind of proposal that NHM was often sent to review. There was nothing unusual therefore about NHM having received it. At the end of a long and exhausting week NHM sat down to read it through once before the weekend. The main job of criticizing it would have to wait until the following week. NHM read through the abstract and the proposed experiments first, and then the background section describing what the applicant had done leading up to the proposed work, and finally the introduction. While he was reading through the introduction Observer flashed a signal to him. Awake it said, you have just read words that you yourself have written and published. Check to see if they have been cited in the references or acknowledged in any way. If they have not, this is a case of plagiarism. NHM immediately began to search through the proposal. No credit had been given for the offending paragraphs. Creator directed, you must now find the published paper from which the material has been taken, make sure it's a word for word copy, and then get in touch with the granting agency. NHM set about the task. He came to the published version of the review that he had sent in preprint form to the applicant. There it is, Observer said, an exact copy. Isn't it sad for him to have done that NHM thought? Sad nothing, Guardian responded, it was a cheap shot that he gave you. Of all the people his proposal might have been sent to for review by the granting agency it came by chance to you and now that you have discovered it you must reveal this unethical act. Fate has touched you both in this matter.

The following Monday NHM phoned the granting agency and spoke to the appropriate program director. NHM told him what he had found and that he did not wish to continue with reviewing the proposal. He then made a formal request that the agency look into the matter. NHM said he would send the proposal back to them along with the relevant reprint of his own published paper from which the material had been taken. He highlighted the appropriate paragraphs in the reprint of his own work and in the grant proposal. No one could miss that they were identical. After the package NHM sent them had been received an official called NHM and said an investigation would be launched, and that the proposal would have to be withdrawn. He said further that many things had been discussed about the incident that would surprise NHM but that he could not discuss them. He went on to explain that he was only temporarily at the granting agency, not a permanent officer there and that perhaps after he returned to his home institution he might be able to give some details. NHM never did hear anything more from him or the agency beyond the fact that the applicant was forced to withdraw the grant proposal. A week or so later the plagiarist called NHM to say how sorry he was for the entire problem. He claimed that a foreign postdoctoral fellow of his who knew nothing about plagiarism had written the proposal. He said and that he had failed to discover the error before submitting the application. Nevertheless he said that he had decided to shut down his research lab permanently as a result of the error. What a horror. A good deed done ends up ruining someone's research career, NHM thought. Consider yourself lucky Guardian warned you just barely escaped being the loser.

NHM thought that would be the last of his contacts with this scientist but it wasn't. Later in the year NHM was seated in a bus waiting to be taken from the airport to his hotel. He had traveled to attend and speak at a large scientific meeting in Chicago. There in the dark like a bad dream appeared the offending microbiologist. He boarded the same bus, saw NHM and came to sit down beside him. They spoke mainly about nothing on the short trip but finally got around to the grant application. NHM told him it was too bad it ended up the way it did. Creator made him point out, however, that had NHM's paper been delayed while in press until after the grant proposal had been submitted, the grant applicant could have claimed that NHM stole the work from his grant application! Guardian applauded Creator for having realized that. NHM hadn't considered that until that very moment. Finally the bus arrived at the hotels and they parted never again to meet or communicate.

NHM got home from Chicago just in time for a meeting that President Koffler called of the search committee that had been trying to fill the soon-to-be-created position of Dean of the College of Science. The committee had failed to attract a candidate before Koffler arrived and it was important to get the process going again. Koffler asked the members of the committee to initiate a new search rather than to consider others who were of lower priority on the first list. He said he would be very grateful to those willing to remain on the committee for another year. NHM agreed to stay on. So did several others. The remainder opted to resign rather than put in the time all over again. Koffler then announced that the search for a new Provost would soon begin and added that he was going to appoint an interim acting Dean for the College of Science in order to get the college started on schedule. His pick for acting Dean was none other than the man who chaired the committee that developed the organizational plan for the new College of Science, and who failed to make the final list in the first search for the founding Dean. No wonder Koffler didn't want the search committee to propose others lower down on their list; the new acting dean was not on that list. In spite of the obvious, the new search committee then consisting of both old and new members started the search process all over again. The careful evaluation of applicants dragged on and on. How stupid could I have been, NHM wondered as he saw the events unfolding and the fact that the now acting dean had made application for the permanent position. In this search process the acting dean's file always made the cut to those still in contention. The new members of the committee were clearly his supporters.

Meanwhile the search committee seeking to fill the Provost's position carried out their work in parallel. The heir apparent to that position, Dr. Paul Rosenblatt, the old administration's "Renaissance man" sank out of contention. He was not going to get the job and he was very bitter. Paul resigned as Dean of the moribund though still alive College of Arts and Sciences at the next department heads' meeting. He said very little other than that he had submitted his resignation, and then he stomped out of the room. Stunned silence followed his exit. It was definitely the high point for drama at the department heads' meetings. Rosenblatt's colleagues represented by the department heads from the humanities, social sciences, and foreign languages went wild. They started shouting their support for Rosenblatt and their hatred of the central administration. Rarely had NHM seen grown people become so agitated and disoriented so quickly. He and the head of the physics department tried to calm things down. Together they attempted to restore decorum, with some success but underneath it all NHM believed the two of them were viewed as part of the power structure that had done in liberal arts in favor of a dominant college of science. In one thing the humanists were correct, separation of sciences from the rest did destroy any semblance of a cultured academic community at Arizona. Scholars who were not scientists became second-rate players in an institution that aspired more to be a research center than anything else.

On the dean's search committee push eventually came to shove. Creator figured out what had been orchestrated. The acting dean was going to become the dean just as the past administration had hoped. It took some time however to get this worked out within the search committee so that it would appear that he was the candidate of choice not just the central administration's pick. In the meantime a new provost was hired. One of his first jobs at Arizona was to finalize the dean of science search and forward a recommendation to Koffler. The search committee was summoned by the provost and seated around a large table in the administration building's conference room. After a short introduction and hello, he went right to business. He asked each committee member to give him their assessment of the candidate and whether or not they wished to see him appointed. One by one every single member of the committee spoke with great praise of the candidate, not just those who were newly appointed to the committee to support his candidacy. When it came NHM's turn the Guardian of his core took charge. Performer just delivered the words. NHM let them know how disappointed he was to see politics dominate reason. He elaborated precisely why he felt the candidate was unqualified. He described his belief that we needed a man of much greater accomplishment and much broader vision. His hope was for a person who knew what was going on beyond the walls of the University campus in Tucson. He told them he thought we deserved a person of stature who was respected in the academic community and who would enhance the reputation of our institution. And most important, he said he thought we ought to have a dean who understood the force of science and discovery and how they fit into culture as a whole. He ended by saying that he felt sorry that his colleagues who had participated in the first search had been pressured to resurrect an applicant who did not make the cut to the top twenty five in their initial evaluation of the candidates. Silence followed NHM's venting. The provost then thanked everyone and they left. After his coronation the reign of the new permanent dean passed without notice as Guardian had anticipated. What NHM had said about him at the provost's meeting must have been communicated to him if not instantly then very shortly after the meeting adjourned. Two other deans, one from the college of medicine, the other from the college of agriculture asked NHM if he would like to transfer into their colleges. NHM thought about doing so but President Koffler blocked any such move. Koffler said we need you in the College of Science, Neil. And that is where he remained trapped to this day in 2007, three central administrations beyond Koffler's tenure.

Although the relationship between NHM and Henry Koffler was cordial there was basically very little the two of them could agree upon. Koffler tried to convince NHM to support his plans for biological sciences at Arizona but his arguments fell upon deaf ears. Finally Koffler invited NHM to lunch so that the two of them could explain their views to one another. To speak privately they went off campus to a famous old historic inn, the Arizona Inn, where the guest log has entries from several United States Presidents. There Koffler revealed that he planned to merge all of the biological science departments into a large single group similar to one that he had administered at Purdue years ago. NHM thought it was a very bad idea and told him why. He asked Koffler if he had studied the history behind the current administrative organization, the reasons that the previous administration had built the current structure, and how it had functioned to meet the objectives that it had been designed to achieve. No he hadn't, Koffler said, and furthermore it became clear that he didn't care about that. NHM attempted to explain to him that when John Schaefer decided to create a research centered university at Arizona he started with an institution that had grown from a teaching college. When NHM arrived there were only two academically recognized programs in place, astronomy and related groups such as optical sciences, and anthropology. The former was located here because of the clear skies and nearby Kitt Peak Observatory, the latter because of interest in southwestern cultures. There was a newly created medical school snatched from the grip of Arizona State University by political power, and a large number of departments providing instruction but little research. The institution was not organized to foster research. There was no vice president for research position, nor any program to encourage the writing of grant applications. The small National Science Foundation Research Grant that NHM brought along with him when he moved to the U of A was the only Federal grant in his host department at that time. Schaefer's idea of creating a Class I Research University required a large change in the culture of the institution. As he began to implement it there was a great deal of stress on the faculty. Insecurity was everywhere and teaching became polarized against research. NHM told him that what Koffler found when he arrived was a structure meant to deal with the transition from a quiet out of the way college into a world-class competitive university.

Koffler wasn't impressed. That's not what I'm interested in he said. Will you support my plan he asked? How can I, NHM answered? It's not the right thing to do at this time. That was not the right thing to have said to him. Henry Koffler became very angry. His face turned bright red. He began to shout and got very agitated. Those in the dining room who knew who Koffler was conspicuously avoided looking over at him. They acted as if they did not hear his outbursts. For a moment NHM thought Koffler might have a heart attack. When he finally calmed down, they left. Koffler drove them back to the campus and told NHM as he left for his laboratory that he would soon be announcing the plan to join the departments together into a single entity. What a fool he must be Creator said but you shouldn't have rebuffed him so definitively. Oh yes you should have, Guardian objected. It will do NHM more harm than good, Creator shot back. That is impossible Guardian told Creator. The kind of "harm" that can be done will never be of significance to NHM.

Word leaked out before Koffler's official announcement of the reorganization was made. There was significant opposition to his plan across the spectrum of political views about how the university should be organized and who should be making decisions about programs. The administration immediately began damage control measures. A series of meetings were arranged at which Koffler spoke to various groups making assurances that the proposed changes would only lead to benefits on their behalf. Never mind that there were finite funds available for which each program would have to compete, Koffler's new plan would benefit everyone. At least he tried to spin it that way. Not everyone believed him. NHM was invited to attend one of Koffler's pep talks meant to calm the campus microbiologists. Members of the medical school department in attendance were very opposed to Koffler's plan and in some ways to him in general. During the Schaefer administration the central administration had fought a tough battle to curtail the autonomy of the medical school administration and many in the medical school were still quite bitter about it. At Koffler's pep talk a heated argument ensued that got way out of hand. The language was anything but refined and the insults became personal rather than honest or focused. It reached the point where finally NHM came to Koffler's defense! "Look, he's the President", NHM said; we have got to give him a chance to do what he can for the institution. NHM's unexpected support of Koffler stunned the group just long enough to defuse the heat of the attack. Koffler seized the opportunity to hastily conclude the meeting and depart.

On the way out he stopped and said, thank you. But nothing really changed either with respect to their relationship, NHM's opinion or the opinions of the microbiologists about Koffler and his proposal. Koffler eventually realized that he would not be able to achieve the sort of reorganization that he had in mind. To save face he scaled down his aspirations to a bare skeleton of the initial plan. Today virtually nothing remains of his reorganization. Koffler eventually retired and became a real estate developer. He built a retirement community in the nearby desert, called the Senior Academy. Last NHM saw of him was in an advertisement for it in which he stands in the desert smiling at the beauty of nature. Hidden behind a new beard and underneath a hat, he is the picture of self-satisfaction and contentment.

In the CDB department NHM went ahead with the hiring of new faculty using authorizations obtained in previous years. The first assistant professor that he hired was Dr. Wah Chiu, a physicist who had perfected state of the art methods to determine the three dimensional structure of biological molecules using low temperature electron microscopy. Wah was internationally known and respected for his achievements. His first faculty meeting was something to never forget. By chance Wah sat down next to one of CDB's senior professors, Dr. Peter Pickens. Pickens was a classically trained zoologist but he had very few research publications to his credit, perhaps three publications in total. Pete was a very tall and vigorous man, about six foot four inches tall, but generally a soft-spoken person. Sitting next to him Wah looked like a midget, but in fact he was a giant in terms of research productivity. Shortly after NHM had gotten the meeting started an argument erupted between Vas Aposhian, the former Department Head and Pete. It had nothing to do with the items on NHM's agenda for the meeting. It had to do with a grievance that persisted from Vas's time as department head. NHM failed to intervene quickly enough in their discourse before their anger escalated to the boiling point. Pickens began to articulate his views by banging on the conference table. The pounding by his long arm and big hand became more and more violent to the point where the table legs looked as if they would give way with each smash he made to its top. Aposhian sensed that it was getting dangerous, jumped up from his chair and tried to run out of the conference room. It was crowded however and to get out he had to trample over people seated along the wall. After his departure everyone sat still in silence. Before NHM could take control Pickens began to reiterate his points, again banging away. Wah sank lower and lower in his seat and drew away from Pickens. Then Pickens himself rose and he too ran for the door stumbling as Aposhian had on the way out. When he was gone, NHM went to the first item of business on his agenda as if nothing had happened. When the meeting was finished he went over to his lab in the Shantz building to do some last minute things on the computer before going home to dinner. Wah called while he was there.

What's up, NHM asked him? Oh, this or that, he said and he chatted about nothing carefully avoiding what was really on his mind. NHM didn't help him get to the point either. Finally NHM said, Wah I have got to go home for dinner, Joan will be mad as hell if I'm as late today as I had been yesterday. Oh one more thing Wah said. Does Pete always act that way at faculty meetings? No NHM said but it does happen. "Well if he does, Wah said, he should take a "trank" (meaning a tranqualizer) before he comes to them." NHM could hardly keep from laughing. Poor Wah had his eyes opened at his very first faculty meeting. Welcome to university life, Wah.

The second professor NHM hired was Dr. Jim Deatherage. He too was a structural biologist, highly trained and respected as a perfectionist. Wah and Jim together with the X-ray crystallographers F.R. Salemme and his wife P. Weber, then in the Chemistry Department, constituted the beginnings of a focused group that NHM believed would eventually push the boundaries of cell and molecular structure beyond mere description. NHM hoped to include an understanding of mechanics and forces, material properties, and regulatory roles of structure in cell function. His plan was to strengthen this core by adding to the group some engineers, some molecular biologists interested in genetic regulation of structure, and some developmental biologists willing to try to understand morphogenesis in terms of these other parameters. With these complimentary strengths NHM thought it might be possible to gain new insight into biological growth and form, form and function and the progressive changes that living things go through during their existence. The last professor that NHM was able to hire as part of this initiative was Dr. Danny Brower, a developmental biologist/geneticist. His recruitment had to be finalized by the dean of the college of science, rather than by NHM himself. NHM passed Brower's file to the dean and said to him, "this is a good man don't screw up bringing him to Arizona". By the time Brower arrived NHM was no longer a department head, and CDB no longer existed. In its place was a molecular and cellular biology department and acting head Dr. Tom Lindell. The funds to build structural biology had already been taken away and used to establish Neurobiology. Wah and Jim transferred to the biochemistry department but found it an inhospitable environment.

Wah left for Baylor where he had a good career in a community that strongly supported structural biology, which included not only Baylor but also other institutions in Houston such as Rice. Jim hung on for a while under conditions in which it was difficult to work. He was denied tenure. He left to become an officer at the National Institutes of Health. Ray Salemme and his wife also left going first to Dupont then to their own company where they became wealthy and had great productivity. Brower stuck with it until he became the Head of MCB. Not too long afterwards he discovered how badly the dean of science was treating him and MCB. He quickly resigned before he had invested too much time as an administrator. In the history of the MCB department and its parent the CDB department we now have a lineage of department head resignations: Aposhian, Mendelson, Lindell, Ward, and Brower. At the time Ward resigned NHM was asked if he would write a poem to mark his transition from Head back to Professor. What he wrote will appear later. When Brower resigned Guardian told NHM that he had to write something for him too. This is what it turned out to be.

Danny Brower's Finest Hours

The head is an important part.
Without one, ...well you're not too smart.
In Science everyone wants to get ahead.
Think of it......If you're not ahead
you're behind.
Never mind.

In Universities it's the other way.
When you're not a head, you're ahead.
Danny Brower was a head.
Now he's way ahead.

Stay ahead Danny!

Free of administration, Brower moved his research program to the Arizona Cancer Center where his focus became the development of his fly genetics system into a model for the study of cell growth regulation and cancer. Guardian gloats over this. You have done the right thing NHM he says in hiring him and also in the poem you have written for him. Now let us hope he has a little bit of luck in his pursuits.

It is informative to understand how Brower became the head of MCB in the first place because it shows how the forces at play and the people themselves shaped events at the University of Arizona. A brief history of MCB will set the stage for the players. Lindell was a professor in the pharmacology department at the College of Medicine. His participation on various committees in the Medical School brought him into contact with people such as Dr. John Law who recognized that Lindell might be an ideal candidate to administer the CDB Department after NHM's departure. Several professors in CDB including Marty Hewlett supported Lindell's candidacy. Marty had by then risen to the rank of Associate Professor with tenure in MCB. He knew Lindell from the days when he was a graduate student in the medical school's biochemistry department and Lindell was a Professor of Physiology. Hewlett and Lindell were linked also through their strong participation in religious organizations on the campus: Hewlett in Catholic groups, Lindell in Episcopal groups. Tom Lindell is a calm quiet person, now a Deacon in the Episcopal Church and a member of the Society of Ordained Scientists. When approached by the dean of science and asked if he would be willing to oversee the CDB/MCB transition during which time the future of MCB would be thought through and a building constructed to house it, Lindell said he would. It was more or less a planning period and a time when a new permanent head would be recruited. Competitive departments such as biochemistry were very much interested in the establishment of MCB because if it became focused in certain ways it could threaten their interests and aspirations. Biochemistry wasn't alone in thinking that molecular biology was their child. So did genetics, biophysics, cell biology, and others. If done correctly MCB would encompass a broad spectrum of people, who would be affiliated with it as joint appointees. Of all the competitors only biochemistry was in a political position to be able to control the birth of MCB and indeed it did.

Biochemistry derived its power through links to the central administration. The UA President, Schaefer was a chemist educated at Illinois, and he had a senior man in his administration, Dr. Herb Carter, also from Illinois who occupied a position created for him called interdisciplinary programs. Carter claimed he came to Arizona because he had heard while working at the National Science Foundation that UA was about to move from the ranks of a second tier institution into that of the most highly rated institutions. Carter sat on most of the University of Arizona's high level committees that dealt with anything pertaining to biological sciences and sciences in general, and he had access to and the ear of the university president. For a while Carter served as acting head of biochemistry when it's original head resigned. He carefully picked the next permanent head, Dr. John Law, and after Law's arrival Carter continued to strongly support the biochemistry program from his position in the central administration. Together Law and Carter garnered so many resources for their biochemistry program that it nearly killed other programs here such as the graduate program in genetics, and the microbiology programs in the colleges of science, medicine and agriculture. Lindell must have sensed the fact that he would have little to do with the development of MCB's programmatic content during his tenure, although he managed to secure commitments for twelve new faculty positions from the central administration. Much of Lindell's efforts must have been focused on the development of the new building, Life Sciences South (LSS), in which MCB would be housed. He did this job exceptionally well. LSS is by far the finest building of five that NHM had occupied at the University of Arizona over a thirty six-year period.

The search process for a permanent head of MCB was significantly influenced by biochemistry. Eventually five finalists were announced, two of whom NHM knew quite well. Each of the five were brought in turn and exposed to the faculty for impressions to be made on both sides. Of the two NHM knew the first was well received. A man founded in biochemical genetics but well into the transition into molecular genetics, his main concerns with coming to Arizona were the quality of students that he would have to work with, and the culture in Tucson. An offer was made to him but he turned it down. The second candidate NHM knew stood no chance of getting an offer. He was a skilled molecular biologist/structural biologist. In NHM's discussions with him the candidate asked NHM to review for him the goals of the program that he had wished to create and why it failed to be implemented. After hearing all about it he told NHM that he would like to develop the very same kind of program. NHM warned him that if he told that to the central administration it would surely diminish his chances of getting an offer. "So what", he replied? "I wouldn't want to come here if I couldn't develop the program that I am interested in." Not only did he tell them his plan but he pointed out to them it was very much the same as the one that NHM had proposed years earlier. It was, in fact, the very plan that was awarded a decision package the funds of which were diverted to form Neurobiology but the candidate didn't know this. Candidate number two, a very accomplished scientist working at the state of the art wasn't what they were looking for at the UA. He did not get an offer. Who did? A man named Dr. Sam Ward was offered the position. Sam's accomplishments were in the areas of worm genetics and development, a mix of cellular and molecular biology. Biochemistry strongly supported his candidacy. When asked why they favored him over the hard core molecular biologists, the most they were willing to say, at least to NHM, was that they thought they would be able to work well with him in developing programs. Sam came, took charge of the new building, and quickly hired three new professors who in turn influenced the bringing of others and pretty soon MCB had the flavor of a Ward Department. Lindell reverted to being a professor in MCB where he developed courses in bioethics and religion. He gave up working as an experimentalist altogether.

Sam Ward must have been given stern warnings about NHM early in his career at Arizona, possibly even before he arrived. He knew about the evaluation that NHM had written of the MMT department and its head, and the way NHM felt about Biochemistry's dominance of other programs. He also knew that NHM was willing to speak out against things that he felt were dishonest, or unethical in nature. Basically he distrusted NHM and thus he marginalized his role in the creation of the new MCB program. NHM paid little attention to Sam and the department as a result of a crisis that developed in his research program, although Ward did not know anything about it at the time. A former friend of NHM's, Dr. Arthur Koch, had attacked NHM's work and tried to discredit it in the strangest way. He got up at a meeting at which they both spoke, immediately after NHM had given his research talk, and made a long speech about how NHM's ideas were all wrong and what it really must be. His comments appeared to be so off base that at first NHM thought it was a joke. But Koch was deadly serious. When NHM described why Koch's suggestions were completely wrong pointing out the known facts that refuted his ideas, Koch simply said, I take back what I said. NHM didn't give it any further thought until several months later when they were both speaking at another meeting. Koch again attacked what NHM had said, this time for a totally different reason. As in the first incident NHM responded by telling him and the audience why his criticisms were totally off base. This time Koch just sat quietly with nothing further to add. NHM knew then that something was terribly wrong but did not realize how much damage Koch could do. Koch had a large following among microbiologists including a group of people he had intimidated into working with him by bullying them with his mathematical skills. These people provided Koch with their data for his use in model building. Then like sheep they simply went along with whatever he told them their data meant. It did not take long for them to join his position although few if any of them knew anything about mathematics or helical growth and macrofibers. Koch's criticisms spread to significant places such as the granting agencies, NSF and NIH, as well as the editorial boards of the microbiology journals such as the Journal of Bacteriology. A program officer at the National Science Foundation told NHM on the phone, "Dr. Mendelson, I hope you're not interested in macrofibers for their own sake rather than what they can tell us about bacterial cell walls". Guardian's blood pressure shot up so NHM responded: That is the most unethical thing I have ever heard from a Program Officer. It is not your position to tell me what I should or should not be interested in. Furthermore, you have just presented a political position originated by Arthur Koch who, for reasons unknown to me, has attacked my work. The woman at NSF responded by having a coughing fit, almost an instant croup, followed by saying, well even we occasionally make mistakes! Performer and Guardian both wished to further discredit this woman, perhaps to raise an ethics complaint against her at the NSF but Creator convinced NHM that it wouldn't be worth the effort. Including it in the story will be good enough he counseled.

To solidify his opposing position Koch wrote a review article for the journal, "Microbiological Reviews", a publication of the American Society for Microbiology. He sent NHM a preprint version of it and asked for his comments. NHM could not believe what he read. Koch's review was filled with misinformation, distortion of facts, missing details, ignorance of essential findings, and attempts to give the impression that he had made the key findings about helical growth. How could this have gotten past the reviewers he wondered? NHM decided not to correct any of Koch's errors for two reasons: first he did not want to get into the position of having to write both Koch's and his own papers. In addition, he believed that by allowing the paper to appear in print then forcing Koch to retract it he could expose Koch's dishonorable intentions as well as the inadequacy of the reviewing process. After Koch's paper appeared NHM sent a very detailed letter pointing out all of the problems in it to the editor, Dr. John L. Ingraham. Later NHM was told that others did too. Ingraham wrote back saying simply that he had forwarded NHM's comments to Koch and would let him know when he had gotten a response.

Months passed with no word from Ingraham. Finally a one-line response arrived that Koch had drafted and Ingraham had agreed would be adequate. NHM did not agree. He pressed Ingraham to force a withdrawl of the entire article. Ingraham refused. NHM then called upon the head of the ASM publications board, Dr. Helen Whitely, who unbeknownst to him was on her deathbed at the time. NHM knew Helen from her work on phages of the Bacilli and she was familiar with his work. He told her the problem with Koch and Ingraham. She promised to encourage Ingraham to force a retraction, and she did. It was nothing as detailed as NHM had hoped to get but it was the best that he could do and he was thankful for having it published. Take a look at it, the "Author's Correction" as it is titled in Microbiological Reviews, Volume 53, No. 2, June 1989: "Biophysics of Bacterial Walls Viewed as Stress-Bearing Fabric. Page 272. Notice that Koch admits to having "garbled Mendelson's point of view...". Why Koch decided one day to attack NHM's work remains a mystery. NHM does have a letter that Koch wrote to the editor of a British Journal, Science Progress (Oxford), in which Koch admits to having started the antagonism, but Koch said nothing about his reason for having done so. The main gist of Koch's letter to the editor was to ask for the opportunity to refute what NHM and his colleague John Thwaites had written about Koch's errors in a paper that was published in Science Progress. The editor turned down Koch's request. In the end Arthur wrote his own book and devoted more than a chapter in it to what he called the two viewpoints, his and NHM's. Koch asked NHM to provide figures for his book! NHM agreed to with the proviso that Koch "use them responsibly". What Koch did was not just based upon a scientific difference of opinion. It went much beyond that. It was closer to an attempt at character assassination and intellectual property theft than a mere controversy. Yet others have told NHM that Koch has great respect for NHM's abilities. Even Creator cannot reconcile these two opposing views. Guardian counsels to ignore Koch's "respect".

The damage that Koch caused unfortunately was not as easily controlled. Koch managed to get himself on the Extramural Grants Program at the National Institutes of Health, and shut down NHM's funding. He was able to do so for the following reasons. The Extramural Grants Program had come to the conclusion that work of a very unusual nature, such as the things that NHM did, could not be properly reviewed and evaluated in the standard manner. When processed through the existing study sections, these unusual proposals never obtained a sufficiently high score to permit their being funded. An alternate mechanism was established. Special study sections were created specifically to deal with a small number of these difficult proposals. The reviewers for these proposals were selected to be the most knowledgeable about the projects for which funds were sought. The responsibility for selecting them was in the hands of a Scientific Review Administrator. In one case the administrator unfortunately made a horrible mistake. He asked Arthur Koch to chair the committee that reviewed NHM's grant application. Koch agreed to serve although he should have excused himself. NHM's proposal was given such a low priority score it was not funded. It was the first such failure for NHM in over twenty five years. Out of funds, NHM lost his research technician and all help. But he was not out of experiments that had to be done and papers to be written. Fortunately he had adequate supplies to be able to continue working by himself, and he had a lot of new findings that needed to be drafted into papers and published. So NHM set about working as quickly as he could to accomplish all the necessary tasks. At precisely this time Dr. Ward informed NHM that he would have to move out of the Shantz building into the new LSS building.

NHM pleaded with Sam to allow him to remain at the Shantz laboratories that he had designed and found to be ideal research quarters; but Sam insisted that he move to the LSS building. NHM asked him why he couldn't just finish out his career in place. Although NHM was determined not to allow Koch to push him into an early retirement in research, he really couldn't be certain that he would be able to overcome Koch's political assault. Ward said there was pressure on him from the college of agriculture to reclaim their space, never mind that it had been rebuilt specifically to NHM's specifications and with the approval of the dean of agriculture. Ward claimed as well that he felt that there had to be some senior faculty in the LSS building then populated almost entirely by the young faculty he had hired. There were other members of the MCB faculty housed outside of the LSS building who were not forced to move, but NHM was. So, everything that could be packed into boxes was. The packing was done by a part-time undergraduate student assistant that he hired and by NHM. They worked day and night to get it all packed by the assigned date of the move. All the large equipment was moved on day one. The next day all the boxes they packed were taken. That morning before the movers arrived NHM decided to dismantle the lab's MilliQ water purification system to make sure it did not get damaged. He carefully disassembled the various components and organized them so re-assembly in LSS would be easy. Then something went terribly wrong. NHM removed a retaining cap and a ball was shot out of the valve beneath it by a geyser of water that rose to the ceiling and began flooding the room. He quickly went beneath the sink to the shut off valve but there was none. No one was there. He shouted "pipe burst" at the top of his lungs, "call a plumber, quick there is a flood in progress". Meanwhile he retrieved the ball and retaining ring and attempted to cap the water stream. The water pressure was too great to be overcome. Attempts to redirect the water stream into the sink failed as well. NHM became more and more frantic. People stuck their heads into the room to view the crisis but offered no help. NHM continued to try to cap the stream. It seemed hopeless but somehow he managed to do it. He was drenched from head to foot. The plumber then arrived with a large wrench. He located the shut off valve and closed it. How did you cap it he asked? NHM told him. He said impossible. The pressure is too great for that to be done, he replied. Yes I guess so, NHM responded and at that precise moment the movers arrived. The first one in saw NHM and said, "hi, how's it going?" What could he say but fine thanks!

Over in the Life Sciences South building there was a problem. The one laboratory assigned to NHM was only large enough for about ten percent of his things. What should I do with the rest of it he asked Dr. Ward? Store it on the second floor he instructed, in the space that will not be finished for a year or so, and he took NHM down to show him where to put things. Into the lab went the very few things needed to continue just what NHM wished to work on at the moment. The rest was piled to nearly ceiling height filling a space about equal to the full volume of two laboratories lacking any benches or furniture in them. In addition to that there were still a lot of things left in the Shantz lab and office that had not gotten packed in time for the movers. It fell to NHM to transport the remains by himself, including all the contents of his Shantz office. He packed it all in boxes over several weeks time, and transported them in his old Volvo Station Wagon one load at a time. There were no appropriate places to park at either building where he could load and unload the boxes. The staff in the MCB office instructed him to park up on the sidewalk near the freight entrance to LSS where utility trucks often did. He did, but of course he was ticketed for parking there. The MCB office staff said they were sorry. In spite of all this he still did not see clearly enough that his life was too difficult. Arizona was not the place for him. Without mentors and the support base needed to be a research scientist/professor in the modern era, perhaps no place could have been the right place for him. He had chosen an inappropriate career. He did not realize early enough in life how important it was to have a sound financial base. Nor did he understand what it was that others had that made it possible for them to always gain the upper hand. This question has plagued him throughout his entire adult life, much as it did Walter in Lorraine Hansberry's famous play, "A Raisin in the Sun". Either NHM was a "born loser" or he didn't know the meaning of success.

The people who knew NHM from his research accomplishments had no idea of what it was like in his working environment and those in his working environment had no idea of where his discoveries stood in relation to those of others. One of NHM's colleagues, Dr. E. Freese, a scientist who worked at the NIH and was well known for his findings concerning the mechanism of chemical mutagenesis, had a different interpretation of things. He felt that it was fortunate that NHM worked in Arizona, far away from the competitive pressure imposed on those working in east coast settings. Ernst believed NHM could only have done what he did in a place where the locals could not detect how different it was from the mainstream. NHM doubted this. The island that was his scientific existence should have been able to function in any setting, he believed but Arizona appeared to be putting this belief to the true test. Was his resolve beginning to weaken? Guardian thought not but Creator was somewhat concerned. Dr. Ward knew nothing of the situation. His concern was to fill the space he was responsible for in the LSS building or lose it to biochemistry and NHM was one way to keep that from happening. Ward also realized that NHM would have to teach something in the MCB program, the other thing he was responsible for as the department head. For most of his career at Arizona NHM taught an advanced microbial genetics course but there appeared to be no future for that in MCB. He relinquished the course to two new faculty members who were hired in the college of agriculture and who pursued molecular genetics approaches to study prokaryotic systems. In the very last year that NHM taught it the students paid him an unexpected honor. To celebrate all of them having gotten through an exhausting term consisting of lectures punctuated weekly with the reading of papers from the primary literature, 18 papers in total each term, NHM scheduled the last class meeting to be a small party. He brought pastries and the like and together they just spent the hour discussing anything the students felt like talking about. Without knowing this would be the last of his teaching the course the students decided to bring some balloons, and to make some party hats by folding newspaper. One such hat was made especially for NHM so he donned it with elan and wore it for the hour. By chance Marty Hewlett passed by the classroom and saw him wearing the balloon hat. How he mustered up a camera may never be known but his efforts led to the picture shown on the next page, which NHM was grateful to have. If you look carefully you will see that by judicious folding of the paper the students managed to arrange for the side of the hat to read, "MAD". How would NHM have ever known that without Hewlett's photo? I guess they appreciated a rigorous course in microbial genetics, he thought.

When Sam Ward realized that NHM did not teach any part of the MCB curriculum he called him in and asked if he could organize a colloquium for graduate students in MCB that would meet weekly and present talks on current papers. He offered a new MCB Assistant Professor to help with it. NHM accepted the responsibility but set new ground rules for how it was to operate. The key thing that he insisted upon was that each graduate student speaker would have to come to see him for a tutoring session about the paper(s) that would be presented and the way in which the talk would be organized. There were two purposes in so doing. The first was to make sure the student knew and understood what the paper was all about and what the data in it showed. The second purpose was to make sure the students could organize the talk well enough so that everyone listening to it could learn something from it. Classical journal clubs as these courses are often called frequently fail badly because no attention is given to the student's communication skills. Once the audience discovers that little or nothing can be learned from such talks they stop coming and the purpose of the course is defeated.

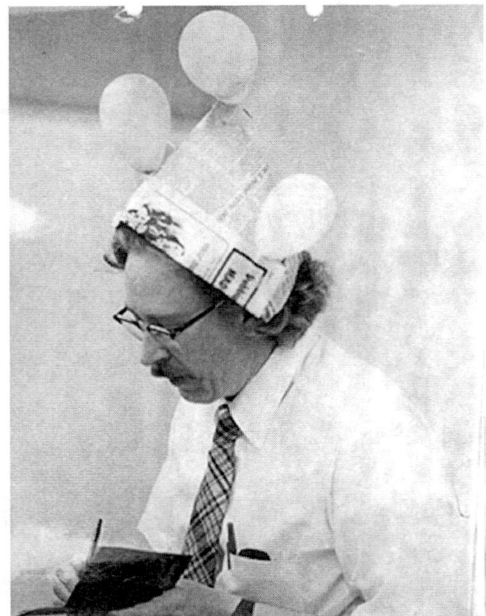

To combat such attrition NHM offered these guidelines.

Three things not to forget on the way to your talk

1. COMMUNICATION SKILLS

All speakers need to have a realistic assessment of their own communications skills in order to use them most effectively. Essential factors include:

a. Knowledge of the material to be presented
b. Being well organized before you get there
c. Planning ahead to make things clear
d. Being confident and relaxed

e. Making sure the material presented is appropriate given the audience and time available

NEVER WASTE THE TIME OF YOUR AUDIENCE.

2. CRITICISM

Criticism is as appropriate in science as it is in literature and the arts. Simple communication of "facts" is of little value. Plan time in your presentation to be critical.

Things to look for are:

a. Whether the new findings are compatible with or contradictory to existent dogma
b. Whether the claims made are justified
c. Whether the data are internally consistent
d. If there are errors in logic or interpretation
e. Whether appropriate controls have been included

3. PERSPECTIVE

Perhaps the most important thing a participant gains by attending the Journal Club is some perspective on the relationship of the new material to things already known. Either the talk itself or the discussion that follows it ought to touch upon:

a. How the current work fits into the bigger picture
b. What more ought to be done
c. Whether there is anything fundamentally new or significant in the work
d. What the half-life of interest in the work is likely to be

The format worked quite well. Soon the course was being touted as the intellectual heart of the graduate program. NHM learned from the tutoring sessions that he conducted that going over new material of this sort with graduate students was an excellent way to assess their abilities and to follow the progress they were making during their graduate years. Of course the students fought back against him and so did the faculty. The control he exercised over them in this course was a bit beyond the current lifestyle of students in America. But they did so in a good-natured way. Here is an example:

NHM thinks they fused his head on another body! The special message pertains to the fact that NHM often made announcements of forthcoming concerts of the Southern Arizona Symphony Orchestra (SASO), and other concerts in which he would be performing, giving them an insiders view of what not to miss. Those who put together the collage never realized that SASO differed from TSO (the Tucson Symphony Orchestra) the main professional orchestra in Tucson! In any event NHM claimed it was sexual harassment to do such a thing as this to an older person and concocted a definition of what the law was, claiming it came from his wife's collection of the Arizona Revised Statutes. Most of them knew that his wife was an Assistant Attorney General so it sounded possible that it might be true! NHM felt that he had to retaliate in some way. When the following appeared at one of the MCB Department research retreats he knew that he must retaliate and it would have to be in a significant manner.

In the picture shown below they mocked the fact that a praiseworthy News and Views article was written about NHM's research in the prestigious journal "Nature". In this case NHM is sure they fused his head on another's body. Those aren't his shoes. What better way could they have shown that they believed it was the scientist who is all twisted up, not only the bacterial fibers?

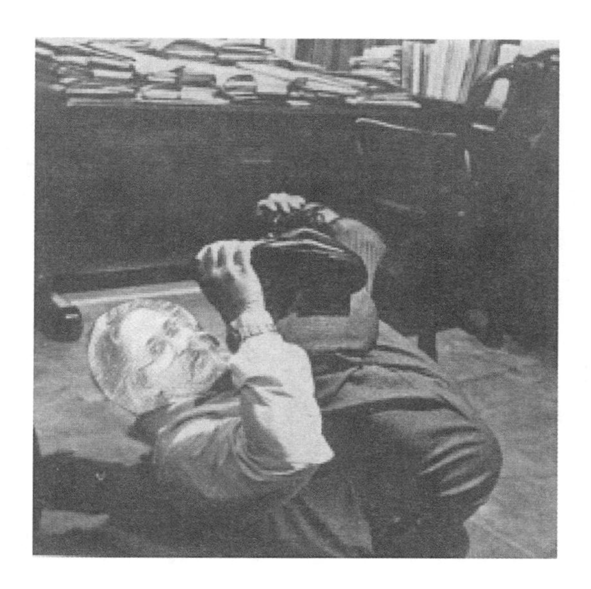

NHM's response came in the first meeting of the course the following term. What he did was this. He scheduled himself to be one of the two speakers at the first meeting in the Spring term. He decided to pick a real topic and a real paper but to design the talk to illustrate everything that a speaker should not do in giving a scientific talk. He found a paper dealing with experiments meant to examine the structural changes that prion proteins undergo when they switch from the benign form into the virulent one that is responsible for mad cow disease. The main idea of the experiments was to use purified versions of the two proteins and to study whether mixing the two would result in converting the harmless form into the dangerous one. The first meeting of the term was always very crowded thus an ideal time to carry out this strange way to teach something. Many little tricks were built into NHM's talk. He planned to talk into the blackboard so no one could hear a thing. He was going to mumble, to stand in front of the overhead projector blocking the image from reaching the screen, and he was going to get things all scrambled up. Then he would reach ridiculous conclusions, emphasize trivia, and run out of time. In addition he totally screwed up the graphics that would be shown. Only one person in the room knew that NHM was going to do something unusual. He was the other person who was also going to present a talk at the meeting that day, a postdoctoral fellow working with one of the new young assistant professors. The postdoc spoke first, gave an excellent talk on a subject of interest to many and set the high standard that we hoped would continue throughout the term. Then it was NHM's turn. The first problem came when he "accidentally crossed in front of the projector and became blinded by its light". When he "recovered" he appeared totally confused. He couldn't find his notes and graphics. He appeared to have forgotten what the talk was going to be about. Then he remembered but it was the wrong topic he remembered. He then gave himself some moments to recover. It all came clear to him then but then something was wrong with the first graphic. It appeared as follows

Clearly it was meant to be an outline of what the talk would include, titled Mixing and Disease, with seven subtopics including, infection, barriers, portals, vectors, symptoms, contagion versus toxin and replication but somehow it was all scrambled up. At first glance it appeared that the graphic had simply been placed on the projector up side down. Up jumped someone from the audience who stormed up to the projector to correct the error. He picked up the graphic turned it up side down and placed it back in a definitive manner. He looked at NHM as if to say you got it upside down you jerk. Then he noticed that it still was not right. So he tried another orientation, a third, nothing worked. Puzzled he put it back down, shrugged his shoulders and went back to his seat. NHM stood as if disoriented. Then he said, “What’s wrong here?” He slowly picked up the graphic and examined it. The clear plastic sheet upon which he had written was in a plastic sleeve. NHM pulled it out from the sleeve and looked at it carefully in search of the answer. Of course there was no answer. NHM had written it so that when placed in any of the four possible orientations it always looked up side down. NHM concluded something must be wrong with the sleeve, a ridiculous idea. He then slyly retrieved another copy of the graphic from beneath the projector and substituted it for the original scrambled one. When NHM placed the new one on the projector everything was fixed. Of course NHM had written the second graphic with the same pen and letter characteristics as the first but correctly this time to make it look as close to the first one as possible. To the audience it appeared that NHM's conclusion that the problem was with the sleeve had been correct. They knew this was impossible, that it was a trick and they laughed, all but the poor man who had been duped by the trick. After the talk NHM went to his lab and told him he was the best straight man that NHM could have hired to carryout the trick. Unfortunately, the victim did not appreciate that. He wrote a letter to the department head demanding that NHM make a public apology to him and to the class for doing such a despicable thing. Of course that never happened, consequently he didn't speak to NHM for years afterwards and he fought against whatever position NHM took on anything during that period. Eventually he did recover and when he learned that NHM was going to retire asked if he could review his work in a Journal Club talk. He did and it was appreciated. Others who had been to NHM's crazy talk later told him what they thought had happened. Some believed NHM had had a stroke. Others caught on soon to what he was doing and loved it. Finally there were those who focused on the real science in the talk and told NHM that something could actually be taught in such a way. From them NHM realized that no matter how bad your communication skills may be there is still a chance that you can transfer some information to others.

Sam Ward assigned two other teaching obligations to NHM. Neither was quite as fortunate as the Journal Club had been. The first was the result of an idea that came from some of the new young professors in MCB. They convinced the department curriculum committee that the department should offer a course to their graduate students in what they called, "Scientific Infrastructure". By this NHM believed they meant the politics of science, and science as a career but he was not sure. Having gotten the idea accepted none of them was willing to develop or teach the course. Would you do it they asked NHM? We need someone who has already developed a career to make it successful, they claimed. What precisely do you think it ought to be about, NHM asked? They could not articulate any detail. NHM went to discuss this with Dr. Ward, and told him what he thought it might be about. That's it, Ward said. It would be great if you could do it. Well here is what NHM did. He chose to examine the period from being hired as an assistant professor through the initial seven-year period to tenure, what it involved and what he believed were the things that had to be accomplished to assure promotion and tenure. The first and most intensive part dealt with grant writing and what it takes to win a research grant. NHM managed to get a group of faculty members to let him use grant applications that they had submitted to various granting agencies. This material was considered confidential and as such was carefully controlled by NHM in terms of its distribution and eventual return. For each proposal NHM also got the reviews that the applications received from the granting agency, the scores the proposals received and whether or not they were funded. This information was assembled into a second confidential book. Each student got a copy of the applications book and was assigned to be the advocate for one or more of the proposals. The students then met as if they were the members of a granting agency study section responsible for evaluating the proposals. The student advocates presented their proposals and gave their opinions of them. This was followed by a discussion in which all the other members of the panel presented their assessments of the application. Finally a priority score was agreed upon. The advocates then wrote a summary statement for the proposal that they presented which was meant to be returned to the applicant along with the agency's position concerning whether it would be funded. Once all of the proposals had been processed in this way NHM distributed the book of actual reviews to the students. Together they then made comparisons between the real reviews and those that the students had written. It was amazing how congruent the two often were. When they were not, the group tried to figure out why the two differed. This exercise was followed by having the graduate students write their own grant applications. Once again the group assumed the identity of a grant panel this time convened to review their own proposals. NHM too evaluated each of their grant proposals. The student's proposals were also ranked much to the dismay of their major professors, whom NHM suspected had spent much time helping them draft their proposals; something that wasn't supposed to happen.

While the students worked outside of class on their proposals, in class NHM examined with them the benchmarks that would have to be met during the initial years as an assistant professor. He talked about the need to have a steady publication output, quality versus quantity of papers published, the need to become known as a specialist in some particular area, the importance of getting to know the leaders in the field, the need to avoid distractions, finding and working with mentors, and just about anything he could think of that came into play during this critical period of career development. NHM looked at what it meant to be denied tenure, issues of ethics, and many things that he knew were important although he himself had been unable to make all of them work for him in his own career. As the things he did in this class became more widely known a group of young professors began to sit in on the course. Some came to discuss their own route to success up to that point in their careers. More than once NHM was astonished by what they had to say and wondered whether they would pass the big tenure hurdle themselves. Other departments also tried to send their students to take the course. The MCB faculty however voted against that, intent on keeping the numbers enrolled low so that the message could definitely be received.

A photographer appeared in the class one day, having been sent by the publishers of the University of Arizona General Catalogue. His goal was to find classes that could be used to illustrate the scholarly environment at Arizona for the biannual catalogue. There they were sitting around the conference table with the grant books open and debating the merits of a proposal. It looked like the real stuff in the Catalogue as if some great wisdom was being transferred from the old to the young right there in Tucson. NHM got a lot of mileage out of that picture. Many in MCB felt that he really didn't teach anything because his courses were not part of the required core in MCB, yet there it was, his class not theirs was the Arizona General Catalogue ideal! That very year in the Scientific Infrastructure course, NHM got the first of many signs that all was not well with the MCB faculty. A young woman assistant professor in MCB asked if she could discuss her career and its many successes. NHM agreed but was a bit nervous before her visit. She was the only member of MCB who was not hired by a competitive process but rather was simply placed in the department by the central administration when her husband was hired to develop a new university program. Would she provide something the students could look up to as a model for their own careers, he wondered? In class she began by describing her graduate education, and how she met her husband, then went on to say that she had chosen her postdoctoral position not on the basis of the science but on the basis of her heart. It was meant, she said, to keep her in contact with her future husband. Her postdoctoral work was uninteresting to her. There was nothing that NHM could say but it turned his stomach. It was an affront to what he was trying to do. In five minutes she transformed an academic environment into a television soap opera. Guardian ranted and raved about it. He chided NHM for having allowed her to speak in his class. He argued that NHM should begin a dialogue to follow up on what she had told them in order to analyze the difference between real world behavior and the ideal of a pure scientist who believes that the dedication to science requires a higher standard of behavior. Creator was very strongly opposed to Guardian's idea. He was of the opinion that those places in the world that would support Guardian's perspective were rapidly disappearing, that the University of Arizona was not one of them and that it would be much wiser for NHM to simply swallow the insult and move on. Performer sulked that he wasn't getting his fair share of the action. NHM sided with Creator but promised Performer that there would be much for him to do and not to worry.

The next young professor to appear in the Scientific Infrastructure course was a man who had just been hired in MCB. He arrived full of enthusiasm and energy. By the time he came to share his impressions of the profession he revealed his disappointment in discovering that he could not meet the demands of teaching, research and service at the same time, as would be required for becoming tenured and developing a long range career. Eventually he gave up his position and his research in order to accept a teaching only position as an instructor. And so it went year in year out until finally it became obvious to NHM and to the rest of the MCB faculty as well that the Scientific Infrastructure course no longer served a useful purpose. The reasons for that were this. Most of the MCB graduate students were no longer planning to find careers as professors. The dominant research mode of MCB's young professors was that of a group effort rather than an individual effort. It was no longer appropriate to emphasize conceptual leadership or the development of new approaches. The rules of the game were not those that NHM had played by. Either someone else would have to redesign the course or it would have to be terminated. No one else wanted to invest any effort in it. It became history and was quickly forgotten.

Sam Ward was not one to give up easily in his quest to have NHM teach something to the student flock that was his responsibility as head of MCB. But what should he teach? As a result of student enrollment numbers and scheduling problems with the young faculty, Sam called NHM in and told him that he needed some help. All the newly hired MCB professors were enjoying an initial year in which there were no teaching obligations, a new perk used to lure candidates to MCB at Arizona, consequently Sam fell short of teachers. Could you join in the team teaching of a senior level undergraduate course in cell biology, he asked NHM. He said that he and Marty Hewlett would each teach one third of the course and wanted NHM to teach the remaining third. What material would I be responsible for, NHM asked? What do you think would be appropriate he asked NHM? How about prokaryotic cell biology? "Why not?" Sam replied. So NHM put together one third of a course dealing with bacterial cells, something that almost never appears in a cell biology program because it is usually claimed by a department of microbiology. NHM knew full well that MCB students would have no specific background for his section but assumed that they would be at the senior level in ancillary subjects such as chemistry, physics, genetics, molecular biology and the like.NHM planned to build upon this base insight into the world of bacterial cell biology. NHM's plan failed. His first inkling that it might not work came when he attended the lectures given by Ward and Hewlett. Although the class was billed as a four hundred level senior course what they were being taught was so elementary he feared it would be redundant with things taught them in high school. It wasn't. The questions asked by the students in class revealed that the basic concepts had escaped them. There were two possibilities: either they had never been taught anything, or by the standards of the academic world they were simply not competitive. Frightened though he was NHM stuck to his goal of starting from scratch and developing everything from within concerning bacteria and their relationship to eucaryotic cells. Once into the lectures it was quite clear that virtually everything he said had never been heard before by the students. Not too long into his section a student asked him where he got the information that he presented to them in a table. NHM explained that he had compiled it over the years from many publications. That wasn't good enough for the student. He wanted the reference to a book where he could find it. Professor NHM explained that it couldn't be found in any single book. Then he said, look what I tell you in lecture are my interpretations of factual information found in the scientific literature. Your job is to go to the sources, find the information and other's interpretations of it, then synthesize your own understanding of these facts. This is a senior level course, he told them, not high school. Do you or do you not know how to go about learning something new? The sad answer was that most of them did not know how to go beyond either repeating back what they heard in a lecture or simply restating something they found in a textbook.

It was a scandal. NHM informed the MCB faculty of his discovery. One-by-one the faculty admitted that they too had experienced the same problem in their classes. They described their disappointment at discovering that our students had not mastered even the most basic ideas. And, they voiced their frustrations with trying to offer advanced topics for which no base existed. The students responded to NHM's challenge by writing letters of complaint to the department head, and by giving him a terrible teaching evaluation. Months to years afterward some of them had gotten into real life situations in biological research. There they discovered that what NHM had tried to teach them was what they needed to know in order to be competitive. Some of them wrote to the department head to tell him how wrong they had been when they had written NHM's teaching evaluations. Neither their impressions nor their evaluations did any damage to his career however. They might have even helped it. Sam Ward decided that it might be better if NHM didn't do any more teaching in the undergraduate program. That had one main advantage; it protected NHM from further discouragement with the academic level of the students. This form of discouragement took its toll, however, on quite a few of the younger faculty members. Some years afterwards NHM did enjoy being nominated and accepted for inclusion in "Who's Who Among America's Teachers". The beauty of being included in this particular Who's Who is that it requires you to be nominated by a student of exceptionally high academic achievement who considers you to be the one teacher who has made the most significant impact on his or her life.

In his role as head of the MCB department Sam Ward knew that it would be difficult to nurture our young faculty through their trial period leading to tenure without some help. He himself had never overseen anyone's promotion and tenure either at Arizona or anywhere else. Although there were in place some department guidelines setting benchmarks that should be met at certain times Ward realized there was more to it than that. He needed someone with experience to oversee the entire process. It was no surprise that he sought help from NHM. Would you please chair the department's Promotion and Tenure (P&T) Committee he asked? NHM had no way to say no. That set into play a yearly routine in which an evaluation was made of every untenured assistant professor in the department. In some years the P & T committee also had to evaluate tenured associate professors who were going to be considered for promotion to professor, sometimes called full professor as if either their heads had reached maximum storage capacity or they could take no more of the job. Some viewed it as if they were filled up to here so to speak and needed something to attest to their endurance. Promotion and tenure along with recruiting were considered the main factors controlling the quality of a university, sort of like the core of an individual. And, like the Guardian of the core in NHM the university guarded these processes with all its strength. Mistakes were not permissible. As a result of this the information that a P & T committee had to gather in order to make evaluations went far beyond the details of only academic achievements. Facts about the lives and aspirations of those being scrutinized often became part of the dialogue. It certainly must have been a humiliating experience for those under the microscope. NHM questioned whether there was no line separating the private affairs of these individuals from the pertinent details their employers had the right to examine? Was academic life so all consuming that nothing could be considered none of the university's business? One-dimensional people and nothing more, is that what professors were supposed to be? The answer is yes, to a large degree that's all they are or at least appear to be when it comes to promotion and tenure.

You might think that an evaluation process of the kind that the P & T committee had to conduct might require some expertise on the part of the evaluators at least in terms of the science being done by the candidates. Shouldn't one professor of molecular and cellular biology be able to assess the quality of others? Wouldn't it be logical for those evaluating a person to read the candidate's scholarly publications before passing judgment on their academic performance? NHM would have thought so but that's not quite how it worked. The MCB process was quite different. MCB candidates put together their own files about themselves. Often they felt compelled to include in them details about circumstances responsible for their failure to achieve this or that, or why they were disappointed with this or that, and how they planned to rectify the problem next year. The character traits of each individual were laid bare to the P & T committee in the hope that it would carry the day. Whether person A was more suited to research or teaching or service became obvious rather quickly. Weaknesses were readily identified. When there were publications included in the files, virtually no one read them. What? Why haven't you read them NHM asked? The pat answer was, well I don't consider myself an expert in what so and so does. At the department level NHM took that to indicate that the MCB faculty were unwilling to educate themselves. NHM had seen similar behavior at departmental level search committee meetings and it puzzled him then. He could accept it at higher levels, for example, when examining the credentials of candidates for the dean of a college but he never understood the lack of confidence (or was it willingness, or competence) to evaluate a candidate's science that he confronted at the department level. Those making decisions about other people preferred to rely upon someone else's opinion about the candidate's science rather than doing their own evaluation, reaching their own conclusions, and deciding matters on the basis of their own opinions. Observer pointed out that there appeared to be a striking similarity between the professors and the students he had encountered in the cell biology course. Perhaps there was parity here between students and faculty. Perhaps he should have known that when he taught his section. Creator agreed pointing out that such similarities between the faculty and the student body are obvious at least in the Ivy League schools. Yes I probably should have paid attention to that, NHM thought. Never Guardian screamed. Never in a million years. It is the obligation of those who know to enlighten others with what they know. A rising tide lifts all ships. There is no tide in a lake Creator argued. All you have to watch out for there is falling out of the boat. Rising tides versus rocking boats was NHM's career at Arizona.

So they read through the files provided them, listened to one another provide gossip about the candidate's this or that, heard the opinions of friends on the committee as well as opponents, then settled in to provide the wisdom of the age to help the poor candidates do better. Again and again they debated not what's right or wrong with the performance but how can we fix it? Was there a way someone could help person B become more organized in the laboratory? Could help be offered in the organization of grant proposals for person C? Why did professor D always get terrible student evaluations for teaching, and could anything be done about it? Was so and so spending too much time on distractions that sounded wonderful but were of little value when tenure had to be decided upon? Much as NHM tried to go along with all of these things and to guide the process he found it distasteful in the extreme. These were supposed to be professors and scientists each with individual strengths and weaknesses yet they were all constrained to meet an institutionally defined uniformity of achievement: as the famous saying goes, "and all the children are above average." Above average in what is what the P & T committee tried to figure out. Above average in what, and, what it was that the university thought "above average" meant? Those were the questions. The answers were something else.

One by one the assistant professors approached the dreaded date of their tenure decision. Had they achieved the sought after perfection? In truth, few if any did. If not, how close had they come to it? Fortunately most came close enough so that a case could be made on their behalf. Others appeared to be in a hopeless position. Some recognized this and left before they were forced to leave. Some just let the chips fall as they may. NHM's job as the chairman of the P & T committee was to get the consensus of the committee about each candidate and then to write a very detailed letter to the department head making our recommendations. These letters were very important documents. They were forwarded with the department head's comments and the full candidate file to the college level and higher into the central administration. At most stops along the way the letter from the P & T committee was the most carefully studied part of the file. Consequently, NHM drafted each letter with extreme care and precision. His goal was to integrate all the facts that were supposed to be the key determinants into a picture of the individual as a whole. He tried to show just how the strengths of each person fit into the academic community and contributed to the excellence of the institution in terms of meeting its objectives. It took him many days to get the polished form of each letter. The members of the committee were then asked to criticize his letter and he incorporated their suggestions but never really changed the overall assessment of the person.

One by one the junior faculty became tenured, and with that gained the privilege to serve on the P & T committee. Guess what? Many of them were hypercritical of those they now sat in judgment of, critical of those who had much stronger positions than they did at a comparable stage of their careers. Lots of dirt came out on the table. Tales of senior professors trying to get untenured ones to work for them on projects in exchange for helping them get published and tenured, tales of sexual harassment, tales of favoritism or rejection based on color, gender, ethnicity, sexual orientation, religion, political connections you name it. Eventually NHM came to question what this great desert university was trying to protect by its tenure decision policy. Who were these people who decided the fate of others on the basis of the offices that they held? Were they in fact able to meet the criteria themselves? After years of spinning our candidates so that they appeared on paper to be unquestionably the best that could be found anywhere, NHM finally became spun out himself and told Sam Ward that he would no longer chair the P & T committee. Ward announced his resignation at the next faculty meeting, thanked NHM for the work he had done and pointed out that his batting average had been one thousand. Everyone that NHM was responsible for evaluating had been promoted and received tenure. The faculty applauded. Those not yet tenured did not look happy. NHM glanced around the table. He was the least happy of them all for there sat the woman who had spoken years earlier about making her career decisions on the basis of heart not head. She was not one whose tenure decision came through NHM's committee. When the time approached for her evaluation she abruptly left the department and moved to another in the medical school. She was quickly tenured there then decided that really she wanted to be back in MCB and was allowed to move back. Later she became the acting head of MCB, a "Distinguished Professor" in the University, and even an associate dean of the college of science. A lovely person who knows how to achieve what she wants. Here we have a success story that shows what you can expect in the real world. Not all routes pass through the same places. No, they certainly do not.

NHM retreated to his laboratory that had finally become established in the LSS building and had gained new funding from NIH. What a wonderful world it was in there. Sheltered indeed but challenging. When you went in and closed your door whether or not something was discovered was solely your responsibility. There were no political factors. Either you did or you didn't. Either you could or you couldn't. Either you liked it or you didn't. Either you were a scientist or you weren't. Nothing could be simpler or more honest. For NHM this was the heart of the matter. This was his root definition of a scientist. You know what you would like to know. You invent the ways to find the answers. You do the experiments. You make discoveries. You study your findings. And in the end you know something in a way others do not. You understand things in a way others do not. You try to inform others about your findings but in some ways it really doesn't matter if they understand or they don't. You understand. You challenge your own assumptions, try to refine your ideas, try to go further, work with what you have achieved. By doing all this you learn to recognize others who live the same lifestyle. Unfortunately your cohorts are a very small part of the total scientific community. You may have to seek hard to find them but they are there and finding them is important. It proves to you that you are not totally isolated in this world. It shows you that others also know of the isolation and the rewards. These are people you feel very comfortable with, people it is easy to communicate with, and people you can put your trust in. NHM has had such treasured colleagues in the far corners of the world and regrets that he wasn't able to spend as much time with them as he would have liked to. But such links that become established endure however distant they may be and often live beyond the lives of the individuals themselves through the relationship of their contributions to yours. They were all like members of a club, the club of those who do or did.

Back in the world of the ordinary, directed by the Arizona Board of Regents (ABOR), the University was instructed to impose an annual self-evaluation of faculty to be used for, among other things, decisions about how salary increases should be awarded when and if funds were available. The process involved filling in forms that left no stone unturned regarding what you had achieved or attempted during the time you were paid to work. In other words the ABOR wanted to know exactly what you did for the money they paid you. Were you worth it, in their opinion? All the forms you filled out went first to the department head, the person whose responsibility it was to make sure you were worth it, and to decide how much reward you might be given for your efforts. Of course when it came to academic achievements the department head never really felt secure in evaluating what each person had written. God forbid someone might have discovered the origin or meaning of life and the department head missed it in reading the self-evaluation. Or perhaps the head would have to evaluate whether such a claim was true. So what would you expect the head to do? Appoint a committee of course to read the files and tell him what the committee thinks it all means. Each year a different group of tenured faculty was selected for this esteemed job, and one was asked to chair the committee. So here we are during the year when NHM had to be the chairman. He read through all of the files and wrote evaluations for each one. So did all the other committee members. When they met the job was to go through each file, discuss each reviewer's conclusions about each file, and come to an agreement about what the committee's recommendation to the head should be for each person. This was done in the standard manner. Then NHM told the others that he had found the group as a whole to be in a very shaky position. He said that if he were the head of a department in which the faculty made the kind of progress that they had just evaluated he would be very concerned about the viability of the program. There were far too many individual problems faced by the faculty to assure a healthy future for the department. Danny Brower who was a member of the ABOR committee said he too had been troubled by the same observation. Others disagreed. No, this is just a short-term perturbation, one argued. But it wasn't. NHM wrote up the committee's report without mention of this larger problem, delivered it to Sam Ward then privately discussed with him his concerns. Sam listened attentively. He then told NHM that Danny Brower had also told him the same thing. Unexpectedly, Sam Ward resigned as the department head the next week. When he announced it to the faculty he said that he had asked Danny Brower if he would consider assuming the responsibility and reported that Brower's response was positive. But it all had to be discussed with the dean first. The dean agreed with the idea if the faculty found it acceptable. I argued strongly for Danny when the faculty met, and it came to pass. Danny Brower was the new head of MCB.

Sam Ward had always been a man with a high profile at Arizona consequently the faculty knew they would have to honor his achievements as their department head. One of our professors asked NHM if he could write a poem to mark the occasion. This is what it came to be:

There once was a man named ward
who thought he would never get bored
as the head of a tribe called MCB
whose members consist of you and me

although we did what we could to steady his ship
the winds must blow and waves must come
the longer the trip the greater the chance
someone will kick you in the pants

you smile and you say thank you sir
I'll ask the dean if he'll concur
you smile and lie to be polite
sometimes its hard to sleep at night

once in a while things go right
most of the time it's the other way
optimists hope to get through the day
there's always tomorrow they like to say

higher ups look down on you
lower downs know you work for them
politics makes it all go
administration is an amorphous show

action is the thing to do
affirmative, confirmative, positive, re- and in-
thinking can help but is secondary
knowledge helps but it's tertiary
inner strength fades with time
chance and necessity rule

one day other things start to look good
things you might do if you could
slowly but surely you come to think
I could enjoy a better stink
if I could get away I would....

you grow as you go and one thing you know
timing is everything so
Bye-bye Sam

A good job done, and better things to come

NHM printed the poem in a box-like frame, had all the faculty sign outside of the box put it in a simple glass frame and presented it to Sam at a dinner in his honor.

During the period when Danny was taking his kicks Sam retreated to his lab, found new interests, spent a sabbatical year in Germany and returned after Danny had also resigned. When Sam got back the interim department head of MCB was the woman who taught us how to find success in life in more ways than one. She was also serving at the time as an interim associate dean in the College and found the latter more to her taste than running the department although things went well in the department during her guard. Nevertheless she announced that she would not continue beyond her first year in the capacity of MCB's leader. Again the department faced an administrative crisis. Who would run things until a "permanent" new head could be found? None other than Sam Ward stepped up to the plate! How can anyone figure out what people will do? What drives them? What might his goals be? Should NHM cancel the poem he wrote for Sam when he stepped down from running MCB the first time? What better things to come did he have in mind? After a new "permanent" head of MCB was found NHM learned the answer. Sam was going to take early retirement and spend half a year at a new home he was building in the beautiful Vermont countryside where he can enjoy the ecology and the culture as well. The dean granted him the right to teach during the half year that he will be in Tucson. Good guys definitely win. Sam passed the MCB torch to Dr. Kathleen Dixon when he made his second exit.

8. Fitting into the real world

The issue of the young professor who spoke about her heart in NHM's Scientific Infrastructure course requires further comment. Why should I, or anyone else for that matter, be appalled at the route she has taken to success, NHM wondered? She is a pleasant and hard working person and she has many rewards to show for her efforts. Like anyone with brains she plays to her strengths and it works for her. Why should that be offensive? Her strategy doesn't harm anyone. No it doesn't. She is generally well liked and definitely not alone in her game. Arizona and comparable institutions harbor populations that are on the same page as she is. That page may in fact be the page for university administrators everywhere. It is not however the same page as that of the scientist as NHM understands it. And that's where the conflict lies. In science it's a matter of what you can do as an individual not the competitive advantage you gain. You test yourself in science rather than leverage the work of others to further your own position. In some ways trying to discover something really new requires a kind of antisocial lifestyle. You must isolate yourself from the world to do the kind of hard thinking and the revolutionary kind of experimentation that it takes to make discovery. It is indeed hard work as NHM tried to convey to the student who sent him the following letter.

12-4-92

Dear Mr. MENDELSON,

I am working on my after School hobbie of writing To The people who have discovered That are Listed in who's who in my Library. I would be grateful if you would write me, on another sheet of paper, Telling me about your discovery of Bacterial MacroFibers I have sent a STAMPED addressed envelope. Did you also invent? I have sent a STAMPED addressed envelope. I hope To be an inventor someday To! Could you give me any advise?

Thank you very much For your Time

Sincerely
Paul Wilson

This is what NHM told him:

December 8, 1992

Mr. Paul Wilson
168 Willow Road
Nanhant, MA 01908

Dear Paul,

Thank you for your letter of December 4th. I discovered bacterial macrofibers completely by accident when looking for something else. It was pure luck. That happens a lot in science when you keep your eyes open and don't ignore unusual things. Before you can discover something though you have got to know what other people already know to make sure you're just not finding something already known. The same for making inventions. So, learn as much as you can in school and by yourself. And don't get discouraged by what people tell you you can't do. Then with a little luck you too will discover or invent something.

I have sent along a copy of a research paper that has a picture of a bacterial macrofiber so you can see what they look like. Best regards.

Yours truly,

Neil H. Mendelson, PhD
Professor

That's the way NHM views it. You have got to know something first, devote yourself to the culture of the scientist, and then you need some luck as well. Clearly it's not a lifestyle everyone is suited to adopt. There are no guarantees. The best of intentions can easily be derailed. But that's the way it is and if all the pieces fit together your life can be a never-ending stream of discovery, improvement, rediscovery and satisfaction. NHM knew of no better way to satiate the innate need to know than to find out things for himself. He attempted to convey these ideas to the students in the Scientific Infrastructure course. At the same time he wanted them to know about the kinds of opposition they would face and how they might be able to overcome some of it with proper planning. All of the strategies that NHM taught however were based upon the ethics of the scientific culture, as he understood it, as he lived it in his own career. What the woman who spoke about following her heart brought to NHM's class was an example that there is more to life and success than his narrowly construed vision of the scientific endeavor.

The route to success taken by this woman and possibly all university administrators is one that appears to NHM to pertain to survival in modern culture. She and her cohort live by a survival instinct. They see position and power as a safety for themselves. They know the value of recognition and rewards compared to discovery and knowledge. They are willing to assume responsibilities for affairs for which they have no formal education or preparation. On the job training is how they learn things; consequently they fail to avail themselves of the mistakes others have made in trying to do the same job. As pragmatists they are forced to make judgments on issues about which they know very little, perhaps nothing in some cases. Observer recalls a period at Arizona years ago when a group of administrators with no knowledge or training in life sciences promoted the idea that a school of biological sciences ought to be created, and one of them even proposed that he should be the head of the school! NHM thought it ludicrous and told him so. Do you think I would be in a position to run a school in your discipline he asked him, to which he responded, no not you but perhaps another biologist could. Fortunately NHM was not alone in objecting to his political efforts to gain control of the biological sciences at Arizona. The aspirant rapidly faded from the scene and achieved no status at Arizona whatsoever. Others however do rise to the top on little more than he had to offer and they become rich in the process. On December 28th, 2004 the New York Times published a graphic entitled, "The Sky's the Limit" showing the annual earnings of Americans who work at different jobs. The average worker brought home $34,065 that year. University Presidents earned $459,643. In 2005 the Chronicle of Higher Education published an update of President's salaries. The figures are now much higher than those reported earlier. Who can blame anyone who knows such things for trying to garner as much money as they can? Economics appears to be something NHM never really respected. This in itself might classify him as hopelessly unrealistic save for the fact that he knew that he didn't know how to play the game.

Imagine the difficulty university presidents must face in having to make decisions about matters beyond their sphere of expertise. NHM's experiences with the six University of Arizona Presidents who had run the University over the past thirty eight years has shown just how difficult it must be to find someone with the knowledge and wisdom required for the job. NHM suspected that none of them had any formal education in the administration of a public university. Observer cannot recall what the first man's background was but the other five were: a chemist, a microbiologist, a teacher of English as a second language, an engineer, and a physicist. And so it goes down through the vice presidents, deans, sub deans, department heads and to anyone who tells others what to do or makes decisions about their work, progress, salary and the like. One must ask why do these people who run the show leave the fields they were trained to pursue, give up the investments they have made in their lives to become experts, and abandon their initial dreams? Was their initial choice an error or a decision made in their youth that no longer meets their needs as adults? Were they perhaps defeated by the system, or possibly overcome with ambition to gain either power or money, or both? Did they have any long-range plans, or was it just an opportunistic adventure for them? Try a little of this, try a little of that and move on. Has anyone ever tried to find out what makes these people tick? Perhaps the Navajos, our neighbors in Arizona have a word that explains it. Read on to discover the word and why it is so appropriate.

The Navajos lead to NHM's neighbor, the woman who works down the hall and around the corner from NHM's office, Ms. Carol Bender. Carol is the director of one of the most successful programs in the life sciences at Arizona, the Undergraduate Biology Research Program (UBRP). She finds lots of money to pay students to do research as part of their undergraduate "experience" to use the vernacular. Her program attracts the most capable students and the professors who can use their talents in their research programs. Carol is a genius at keeping these kids interested and connected to her program throughout their studies and even beyond that into the real world. NHM often teases her that the whole thing is held together with pizza. Carol's UBRP program began publishing a newsletter some years ago that has become a truly sought after thing filled with interesting details about people and opportunities, events that happen to UBRP students while on their foreign research visits, at meetings, or other academic events. Last year an incredible article appeared in her journal describing how a Navajo UBRP student working in Paris at the Pasteur Institute had the opportunity to name a newly discovered gene that governs mouse behavior. It was indeed fascinating reading. NHM couldn't pass up the opportunity to comment on it to Carol. She asked him if he would consent to write a letter about it that she could publish in her newsletter. When NHM agreed to write something he had no idea that this is what it would end up.

Letter to the Editor

(NOTE: The article by ***Nanibaa' Garrison*** in the November 2002 issue of *The Gazette* resonated with ***UBRP faculty sponsor Dr. Neil Mendelson*** who wrote to us. We have decided to take this opportunity to introduce a new "Letters to the Editor" column to the Gazette. Please feel free to send us your reactions to Gazette articles.) Now for Dr. Mendelson's letter...

November 3, 2002

Editor
The Gazette, UBRP
Life Sciences South
University of Arizona
Tucson, AZ

To the Editor:

Isn't it incredible the way information flows around the world. A UBRP student from the Navajo Nation goes to Paris, to Institut Pasteur, the very place I was a Visiting Professor in the 1970's, is given the opportunity to name a newly discovered mouse mutation, and accepts her mother's suggestion to call the gene, "binahoomas". Binahoomas, meaning something is wrong with an individual but who knows what?!

(All of this is reported in the Gazette volume 13, issue 11, November, 2002). I have been searching for this word forever. My world has been full of binahoomases (binahoooma?) yet I never knew what to call them. They are everywhere I have worked as a scientist: in Europe as well as America. Many of them work at the granting agencies and on editorial boards of scientific journals. In laboratories they can be found on both sides of the lab bench, not just in the experimental model organisms. At scientific meetings, wow, see hundreds at one time. The human form is not underrepresented at Universities. Not surprising the faculty is full of them, not surprising because after all many of those who go to scientific meetings are also faculty back at home. Binahoomas strikes close to the heart of experimental science. Genetics was founded on studying binhoomas, how it passes from parent to progeny, and what causes it. All creative work is usually considered a form of binahoomas, at least until it becomes recognized as what makes the world interesting. Thank you Mrs. Martha Austin-Garrison.

N.H. Mendelson, PhD
Professor

Could the Navajo word, Binahoomas, explain why perfectly good professors become university administrators? Or, could perhaps the gene called Binahoomas discovered in the mouse at Institute Pasteur also exist in humans and that's the answer?!

Rather than give up science as a profession as many others do when they believe they have done about as much as they can, or become more interested in doing something else, NHM retained a strong enthusiasm for research. Pushing the boundary of knowledge never lost its allure. He did realize however that the harder you push the greater the burden you create for yourself to answer all the questions you have raised. Something eventually has to be given up in order for something else to be pursued. In his case it was the study of genetic regulation of DNA replication, cell cycle control, and genes that govern bacterial cell wall polymers that had to be sacrificed. He had invested twelve or so years in that and had a lot to show for his effort. But, to focus on helical growth and macrofibers he knew he would have to relegate earlier questions to the back burner, perhaps never to return to them. From his perspective it would be a definite move ahead. At least he thought so. No one knew anything about this new stuff he had discovered. Good. He could start with a clean slate. The focus would have to shift from genetics to mechanics, engineering, perhaps physics and math. Good. He always liked those things. But your foundation in these is very weak, Creator warned, and remember your principle of not going where you don't belong if you can avoid it. Yes NHM told him, I hadn't forgotten that but I believe I might be able to speak the language of these other fields just well enough to get the attention of the true experts and to get some help from them. Thus was born the objective of NHM's major research efforts: i. find the unique things about this new discovery of helical growth and macrofiber self-assembly, ii.identify that part of the biology that is really in the domain of the physicists and engineers, and iii. do as much research on macrofibers as possible using physics and engineering approaches, rudimentary as it might be, to raise their interests. At the same time iv. point out to the biologists why it is necessary to bring these alternate approaches to bear upon the analysis of living things. NHM convinced himself that virtually everything known about cells had to do with their chemistry and informational macromolecules. Virtually nothing was known about their physics. He believed that the material properties of cellular materials had to be significant perhaps even in regulation of cellular processes. There must be ramifications of forces acting in cells, forces that arise as a result of cellular activities such as growth that feedback into the networks that control cells, he reasoned. Some understanding of how cells seem to defeat the second law of thermodynamics and achieve their amazing degree of order and organization might emerge from investigations of helical growth and macrofiber self assembly, he thought. It never occurred to him at the time that things he learned from the biology might provide clues to issues not yet resolved in physics or math as did in fact happen (see Carol Potera's article, "Physics, biology meet in self-assembling bacterial fibers" in *Science*, Volume 276, pages 1499-1500, 1997). Yes indeed, the elasticity model developed by NHM's mathematics collaborators Michael Tabor and his associate Isaac Klapper in response to bacterial supercoiling behavior appears to be able to help explain sunspots! How could anyone have foreseen that?

One of the beauties of discovery is that it can take you in surprising directions. People and ideas appear spontaneously with many different perspectives. A new discovery generates an enthusiasm not just from young people like Paul, who wrote to NHM, but across a broad spectrum of people. Each sees something in a new discovery that gives them an idea about something they have been thinking about. Perhaps it's an answer to a plaguing question, or in line with something they had found. A former professor from the University of Washington, ninety three years old and living in a nursing home wrote to tell NHM how he was reminded of issues in his freshman undergraduate chemistry course that he believed NHM's findings might have solved. The director of research and development in a company trying to find ecologically acceptable ways to solve some industrial problems scratched a letter while flying to Brazil to see if NHM thought it feasible to use his system to meet one of his goals. A man from upstate New York, an amateur scientist with quite a bit of engineering know-how, sent NHM a copy of his small book, "How to Create Life". In it he described experiments he did involving the electrolysis of spring water that lead to the deposition of material on one of the electrodes. The structures he found resembled living and fossil forms of invertebrates. He shared his discoveries with NHM because he thought it might link his work on mineralized bacterial fibers to ideas about the origin of life. He was not alone in this. A professor working on structures found in rocks that he interpreted as micro-fossils, but which others thought might only be inorganic mineral structures also wrote to see whether mineralized bacterial fibers (that NHM called bionites) might be related to his findings. Finally Observer must mention how new discovery can necessitate calling the bomb squad.

There was an era when letter bombs were being sent to university professors by a disgruntled mathematician, known, before he was caught, as the "Unibomber". As a safety precaution the Postal Service developed a set of guidelines detailing what to do if an unexpected package arrived. It consisted of a list of ten danger signs and urged that the police be called when in doubt about a suspicious parcel. A package arrived for NHM that had seven of the ten danger signs. He called the campus police. An officer arrived and together they went over the significant factors. It was a large gray envelope the contents of which appeared to be a mixture of different items, none of which matched anything NHM was expecting to receive from anyone. A prominent cylindrical item could be seen and felt lying near the top of the contents. Neither of them could tell what it was. The police officer realized it was about the size of a blasting cap. He conferred with his boss and they agreed that we ought to get the Tucson bomb squad to take a look at it. Ten minutes later two detectives arrived carrying a small portable X-ray device. In the hall outside NHM's office they propped up the envelope, loaded a film cartridge not unlike that used to take a chest X-ray, cleared the corridor of people, and took their picture. A copy of the X-ray is shown below.

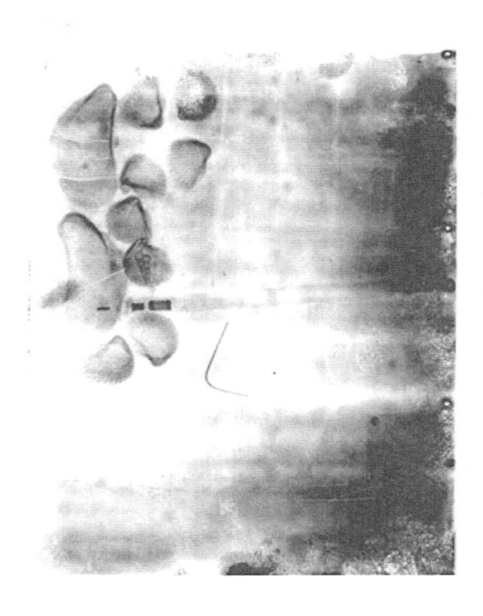

Lying horizontally across the middle of the package you can see that the "blasting cap" was in fact a ballpoint pen. To its left are a variety of seashells. Below it is a bent hatpin. Oh yes, one of the police offices said when he saw the picture; I'll bet I know who sent this. We have seen other packages of hers. It's not dangerous you can open it. What a collection it contained. A hand scribbled note saying that she had seen a talk of NHM's and was very impressed. There were seeds of all kinds. There were credit card receipt slips indicating purchases made in Germany. There was indeed enough so that the identity of the sender could be traced if you wanted to. NHM didn't. He simply put it into his collection of similar bizarre packages eventually to go into the archive that is collecting all his papers, books and the like. This present was the first of three packages that came to NHM, all of which might have been bombs. None were, fortunately.

Here is what the other two packages contained. One had in it an unusual mineral of unknown composition sent from somewhere in New Mexico. It was a slab about 5 inches by 4 inches by 1/2 inch and quite heavy. God knows what it is, or what the message was supposed to be. The package bore neither a letter nor a name. The other package was equally difficult to understand.

It consisted of a cloth bag about the size of a large telephone book. The bag was sewn closed, addressed to NHM and covered with postage from Pakistan. It was quite heavy but didn't feel as if it were a single book. It could have been a group of folders. NHM could not identify from whom it came. Some marking suggested that it might have been posted in Islamabad. What could it be? NHM did not know anybody in Islamabad. True he had had a technician from Pakistan when he worked in London thirty five years earlier (Fahim) but attempts to make contact with him over the years had always failed. What the hell should I do, NHM wondered? Finally he decided to carefully open the package in his laboratory shielding himself from it with a file cabinet acting as a barrier. Its contents were a total surprise. The head of a biology department at Quaid-I-Azam University had sent him the files of applicants for a senior appointment in his department and asked if NHM would evaluate them. Of the group one was a person who had worked in the laboratory of a friend of NHM's in Cambridge although NHM did not remember ever having met him when he was working there. All of the candidates were in fact highly qualified scientists. Most of them held positions in England but wanted to return to Pakistan. NHM called the State Department to find out whether it was permissible for him to fulfill the request. He knew there were some restrictions based upon nuclear non-proliferation issues and wanted to be sure that evaluating potential professors was not a prohibited activity. It was not, so he went ahead and reviewed all the files. In the end he provided a report with details about the strengths and weaknesses of each applicant. He did not however rank the candidates. That he believed they had to do given the needs and aspirations of their department. The senders accepted NHM's work and thanked him for it. To this day he has no idea why or how they chose to ask him to evaluate their candidates. Something doesn't seem right about this. No, NHM's former technician hadn't risen to be the head of Quiad-I-Azam University. Why then would the leading University in a Muslim country seek the advice of an American Jewish scientist? Given current day hatred of Americans and Jews by the Muslim world NHM seriously doubts such a request could ever happen again. Creator never even muttered a thought about any of this but Performer was perturbed. There is danger in having made achievements, Performer warned. Why don't we just fade into the background for a while until it becomes a little safer, he asked? Because it never will, Creator chimed in. We need to move ahead but also to remain aware.

It appears that once your cover is blown, be it because you have made an impressive discovery or done a good job in this or that, or whatever, count on work. Help an editor review a manuscript and pretty soon you're on the editorial board. The same holds for reviewing grant proposals. It all can start in the most logical way. Someone you know and whose work you value realizes you would be a good candidate to read and criticize his or her papers, applications, or even career progress. When you are asked to perform the service you feel it is a professional responsibility and undertake the assignment. Do your job well and word spreads. Here is a good example. A request came to review the grant application of a respected colleague in Australia and NHM agreed to do it. It was an excellent proposal as anticipated, consequently, there were a lot of good things he could say about it. Although the identity of the reviewer is kept confidential NHM realized that on the basis of what he wrote it would not be difficult for the applicant to know who he was. This meant that things NHM disagreed with had to be written in a very cautious way so as not to destroy their professional relationship. Two years later another grant proposal from the same individual was on his table. The following year the Australian Research Council requested NHM's services. Within six months the government of New Zealand appointed NHM to their reviewing council. Not long afterwards the Australian Research Council College of Experts nominated him as an "expert of international standing" and asked for much more extensive reviews. The development of the Internet made it impossible to escape from these duties. No paper is sent anymore in this process. To read a seventy-page proposal that has been assigned to you for evaluation you do so on your computer screen or download it and print the whole thing yourself. You write your reviews on the computer and upload them to a secure server of theirs via the "Internet". In recent years NHM has spent many an hour trying to do so from his farm in the Siskiyou mountains of southern Oregon where the maximum baud rate he can get from a dialup service is never above twenty six kbs. Frustrating, as that might be it is still less time consuming than having to travel to an agency in order to debate the proposal.

The National Institutes of Health and the National Science Foundation make you come to them in the Washington, D.C. area to complete a reviewing process. So does the Educational Testing Service (ETS) in Princeton, New Jersey, where examinations such as the Graduate Record Exams are produced and administered. When the ETS realized that a new advanced subject test was needed in modern biology they asked about twenty five people to draw up a list of the areas that had to be examined. Why they picked NHM he did not know. He accepted the invitation in order to press for the inclusion of topics that dealt with microbes, a subject that was almost totally ignored in their previous exams. NHM believed that if the ETS included such material and made known the fact that it would be on the advanced subject matter test, that would feed back to the universities and colleges and encourage them to include at least some microbiology component in their undergraduate biology programs. ETS eventually did add the topic but whether it really had any impact on what students are expected to learn in most biology programs isn't clear. Not many students wish to study the sciences or engineering in America at the moment. The dominant position the United States held in these disciplines throughout NHM's career is rapidly eroding (see, for example, the article by William Broad in the May 3, 2004 issue of the New York Times). The entire culture in the United States appears to be in transition. Where it ends up is not yet predictable. The direction in which it is headed however has been obvious to NHM for some time. It does not seem possible that the rest of the professorate and other scholars in the US could have missed noticing the trend. It does not bode well for the future of their professions.

Scientific politics is another matter. It's future seems promising. NHM learned a lot about how it operates in this country from participating on reviewing panels. Two cases at the National Science Foundation illustrate what it's all about. The first incident dealt with a proposal that really didn't fit in with all the others being evaluated by a panel on which NHM served. It was a request for funds to support the establishment of a small collection of mutants. NHM was aware of a very large and comprehensive collection that he knew would keep such isolates and asked why the people who submitted the proposal didn't simply send their isolates to the existing culture collection. Oh don't you know, he was asked by one of the other panel members, the head of the collection you are talking about is going to retire and no plans have been made to assure that his collection will continue on. No I hadn't heard that, NHM responded; well then this proposal makes sense. NHM voted in favor of funding it. Several months later the about to retire professor appeared at the University of Arizona and since NHM had known him for many years those who invited him arranged for the two to meet. NHM was then the Head of CDB so the visitor was escorted to the department head's office. What are you planning to do in your retirement NHM asked him? Why do you ask Neil, he said, and NHM told him. That was a total lie he said. I have no plans to retire. In addition, before I moved to my current institution I had made arrangements for the collection to continue on with their support. Although panel members are sworn in before service at the National Science Foundation, none of it apparently applies to honesty. That is not part of scientific politics.

Case number two shows just how complex it can get and how difficult it can be to know what you might encounter on a panel. At Arizona a program that NHM initiated in conjunction with the head of Applied Math received a highly sought after grant from NSF to train graduate students to work at the interface between math and biology. NHM felt compelled therefore to accept an invitation to participate in a panel to judge other such interdisciplinary projects. Once he agreed the NSF program managers sent him the group of proposals that would be his responsiblity, that is those for which he would be the advocate, lead the discussion about them and respond to criticisms from other members of the panel. Included in these was a proposal NHM never thought likely to be submitted for such a grant. It was from the foremost group in the country focused on training students to work at the math/biology interface, a group that set the standard for the field and had fifteen years of outstanding success. NHM knew the key people in this program and was very happy to have been selected to speak for them. Then at the panel it turned rather ugly. One position the opponents took was based on the idea that these people didn't need any new money, which in fact was not the case. Another faulted them because they had nothing new beyond what they had already achieved in their proposal, and a third faction argued that their minority access outreach had been dismal and thus they should be disqualified. That was too much for NHM. He knew precisely how their program had almost killed itself trying to find qualified minority participants. The pool in math is very small, yet they had made some headway. NHM argued very strongly that no program at the level of theirs should be disqualified on the basis of a social objective. When he told them the sign above the door that he came through read the National Science Foundation, not the National Social Foundation, he thought that would be it for this proposal. Observer scolded, you might have just sunk this proposal. Some of the panel members did become inflamed, but the final vote enabled the proposal to advance to the second round of evaluations which would be done by a different panel with other members. NHM's effort kept a worthy proposal alive. NHM believed the NSF program managers knew that this proposal would be a difficult one to get approved and selected him to do the job. Before the meeting it never occurred to NHM that the proposal would be hated on the basis of the applicant program's success. The following year NHM learned that the proposal did not make it through the second round of evaluation and was in the process of downsizing. Could I have been responsible for its demise by the way I forced it through the earlier panel NHM wondered? Perhaps yes. NHM's colleague in math at Arizona assured him however that he wasn't. It appears that a new set of program directors came into power at NSF at the time of the second panel who were just not interested in fostering bio/math initiatives. NHM wished he could have been on the second panel to meet them!

After a long and difficult devotion to macrofibers, devotion to finding out their secrets, to finding money to support their study and to letting the world know how fantastic they are, NHM reached a point at which he believed he had a pretty good idea of how the whole thing worked. Bacterial cells contain information in their growth plan that endows them with the potential to form highly ordered three-dimensional multicellular structures, and they pass this information to their progeny. Everything boils down to simply controlling the twisting of the cell cylinder as it elongates during growth. Beyond that, the rules of physics take over. Constraints result in forces; cell materials deform in response to the forces giving rise to cell shape. Self-assembly is inevitable given the behavior of twisted elastic filaments. What a neat story. Growth provides the cell materials, growth provides the forces, the cell materials respond to the forces, the geometry of growth dictates the shape obtained, and self-assembly is possible because all the cells in the multicellular structure do the same thing as they grow. The only really mysterious part was how negative twist fit into the story. Why is it that macrofibers always supercoil as if they are under wound not over wound? Try as he might NHM couldn't find the answer to that from experimentation. Even theoretical model building failed to provide any insight although he did stumble onto a most unexpected thing from these exercises.

NHM found a new timing mechanism based upon the building of structure that could provide a unique way to link form and time during the growth and division of a single cell. A cell could use this clock to know both where and when an event should take place, not just that it had to do A before B before C, or that it had to divide in half. Everything could be expressed in a single equation containing three variables. Each timing event had a fixed start, passed through an ordered series of intermediates and reached a natural limit; just what one needs to time a single cell cycle. This mechanism which NHM called the, "helix clock model" arose from the study of rules governing the packing geometry of strings wrapped around the surface of a cylinder and the way in which their orientation changed as new strings were added to those already in place on the surface. The behavior of these hypothetical strings could account for regulation of all the central cell cycle processes. It was a powerful model, perhaps too perfect for the ways cells might actually work, but that has yet to be determined. As a disclaimer when NHM gave talks about the helix clock model he would always end by saying that, "if cells really don't work that way you might say there was a design defect in them!" Those who realized what this comment really meant were of two schools. The first had to laugh at the idea that NHM considered a construction of his own mind superior to the lessons of selection and evolution. The second were deeply angry at his insinuation that he knew better than the great power that created life and thus cells in the first place. Neither camp had any ideas however about negative twist.

At the University of Arizona word got around that NHM had developed a new theory about the way cells can time their growth-related cell-division processes. Cancer biologists are very interested in such ideas because they might provide strategies for regulating uncontrolled growth in tumors. NHM wasn't surprised therefore to receive an invitation to talk about his work at the Medical School. But the invitation didn't come from the Cancer Biology Program (which years earlier he had helped organize). It came from the Department of Anatomy, a Department that later expanded its focus to include cell biology as well as anatomy. NHM knew many people affiliated with the Anatomy Department, including a former student of his who remained in Arizona after finishing her PhD and so, although his calendar was very full, he accepted their invitation to give a seminar about his new theory. Observer feels honor bound to tell the truth here. Even if NHM hadn't known a soul over there he could never have missed the opportunity to speak in such a setting. Human anatomy brought old Dracula movies to mind although he knew full well how ridiculous that was. Creator immediately started working on a way to use the opportunity for more than simply showing off his beautiful research findings (as if that wouldn't have been enough!). Just by chance the seminar date was scheduled during the week of Halloween. That coupled with a rumor that came to NHM from friends in the Medical School gave Creator all that he needed. His plan of action had consequences much beyond what he had invisioned. Unfortunately.

NHM had heard from his sources that the graduate students in Anatomy had taken to betting on whether or not he would wear his black cowboy hat while speaking! The day NHM arrived in Arizona he purchased a large cowboy hat that soon became a hallmark of the seasons. During warm months he wore a straw hat; during cold months it changed to a black felt hat. People watched for the change as if it were a requirement of the seasons. Moreover they couldn't help commenting if the transition didn't come as soon as they thought it should have. One spring a representative of the graduate students came to ask why the straw hat hadn't yet appeared. NHM explained that his old hat had been ruined on a flight back from Paris and that he hadn't had the time to replace it although spring had certainly arrived. None of them knew that he started wearing a hat in the first place because his sister had suffered a malignant melanoma at age eighteen. She managed to survive it. NHM knew that Tucson had one of the highest rates of melanoma in the United States and that he would be a prime candidate given his fair complexion and genetic history. On the day of NHM's Anatomy seminar it was black hat weather. His plan was to keep the hat on during the talk to favor the minority bettors. But there is more to it than that.

Creator had come up with an outlandish idea. All his partners bought it. Performer purchased a rubber cap that when worn over your hair made you appear to be bald. On top of it he glued a large plastic nose meant to be part of a Halloween costume. All of it fit nicely under the black hat without disturbing his longish hair that normally stuck out from beneath his hat. NHM took the cap and nosepiece with him in a bag and went to the restroom before the talk to set it all up under the hat. To his good fortune a trusted friend, Dr. Harris Bernstein, happened to be in there when he entered. Harris agreed to help get it all together. It fit perfectly. Harris made sure that none of it could be seen or detected by the audience. Then off NHM went to give his talk. He met the seminar organizer and gave her his slides, spoke to colleagues before getting started and waited for the introduction. When he rose to speak it appeared as if he had simply forgotten to take off his hat. It didn't take long for the audience to become focused on the science not the hat. And so it went for five or ten minutes. Then, having completed the outline of what he was going to show them and saying why it was a fantastic finding, he stopped as if he just realized he had his hat on. He touched the brim with his hand and glanced around as if he was rather embarrassed. No one would make eye contact with him. Not a sound followed in a moment of suspended animation. Then NHM said: "Wait a minute". All eyes focused directly on him then. "I heard about the bet", he said, "and indeed sometimes I do give talks with my hat on but for this talk, OK I'll take it off", and he did. Wild laughter followed. Someone in the audience had a camera and a flash went off. When the laughter died down NHM said simply; no one's perfect. That too triggered laughter. A professor ran up to offer him a yarmulke from his pocket. For what he will never know, nor its purpose. The professor wasn't Jewish! Then another professor rushed out the back door. NHM then settled into his talk and enjoyed the discussion afterwards. NHM's opportunity to tell the anatomists about human anatomy was not wasted. But it wasn't all just laughs either. The professor who ran out was Jay Angevine, a senior neuro-anatomatist. The retina in his right eye had detached while laughing and he ran down to the ophthalmologists to have it reattached. Jokes can be dangerous. Fortunately he regained full use of the injured eye and continued his career without interruption.

Getting back to the nagging issue of what was responsible for negative supercoiling NHM's long-term collaborator John Thwaites and he had many debates about what might lead to it but the conclusions they drew as possibilities never really satisfied NHM. Thwaites, a fellow in Gonville and Caius College in Cambridge worked in a beautiful setting there as well as in the University's Engineering Department. Although an engineer, his main tools were mathematics, hard thinking and working things out with pencil and paper. To this day he has no use for computers, email and the like. If they were not working face to face in either Cambridge or Tucson Thwaites main means of communication was the fax machine. Rather than put one in his lab or office in Arizona NHM relied upon the ones in the MCB department office, as did most of the other faculty. In response to Thwaites' faxes NHM usually called him back at his home but sometimes he had to fax back material or FedEx it to him. On one such occasion NHM tried unsuccessfully to establish a fax connection with the machine at Caius College. There was so much distraction and noise in the MCB office NHM kept dialing the long string of numbers incorrectly. Finally he tried to quiet things down. The main problem was a discussion between two secretaries who sat in different rooms but in sight of one another. One of them was trying to arrange information from potential foreign graduate students into a format required for visa purposes by the Immigration and Naturalization Service. Her problem was an applicant from Bangladesh. What should I do she kept asking her colleague; look he hasn't filled out all these parts of the application. Why do all the Bangladeshi's do this? How can I ever get their visa's approved? To which the other replied, I don't know what to do. If it's denied, it's their fault. That wasn't good enough for the first secretary. I'm going to just make some of this stuff up she said. No, I wouldn't if I were you the other said. Why not? You'll get in trouble. No I won't. I don't care. How can I do my job?

Finally NHM said Ladies. Ladies. The trouble with Bangladesh is ....diarrhea. Silence followed. Then one said, diarrhea? Yes diarrhea. Again not a word came from either of them. Look, NHM said it's simple. The problem for most of our Bangladeshi applicants is that they have to use the paper before they can finish filling it in. In the silence that followed NHM's fax went through. He hurried to the elevator while the silence lasted and returned to his office. There laughter overtook him. What a terrible thing to have said. Oh well. Observer took responsibility for the whole thing. He had sent Creator a fact he once learned from Dr. Stan Falkow, a microbiologist friend of NHM's who actually worked on enteropathogens that cause diarrhea and worse. Falkow talked about the very high incidence of serious diarrhea in Bangladesh. That's all Creator needed to concoct the rest. It just fell into place that day.

The laugh must have cleared NHM's mind because in the next few moments he realized a rather important thing. In Paris when he first developed the ways to culture macrofibers NHM often grew them in test tubes rather than Petri dishes. After overnight growth when he picked up the tubes to examine them the fibers were always on the bottom. Could it be that fibers are denser that the growth medium and settle to the floor? If so touching the floor might block twisting motions and induce supercoiling. Could we have overlooked this in all our work he thought? All our models made the assumption that fibers were suspended in the bulk phase of the growth solution. I better take a look at this NHM thought hurrying to the lab to see how he might rig up a way to obtain a side view of growing fibers and their position relative to the floor of the growth chamber. Petri dishes were not suitable because of their curved vertical walls. Nothing in the lab was suitable, so NHM designed a small chamber built of glass microscope slides glued together with the adhesive used in fish tank aquaria. By week's end he had some leak proof chambers in which fibers could grow but no optical system with which to examine them. Some salvaged old microscope tubes and bodies were found, and with optics from other microscopes used in the conventional manner the whole thing could be plugged into NHM's system of producing time-lapse video films. The results were startling.

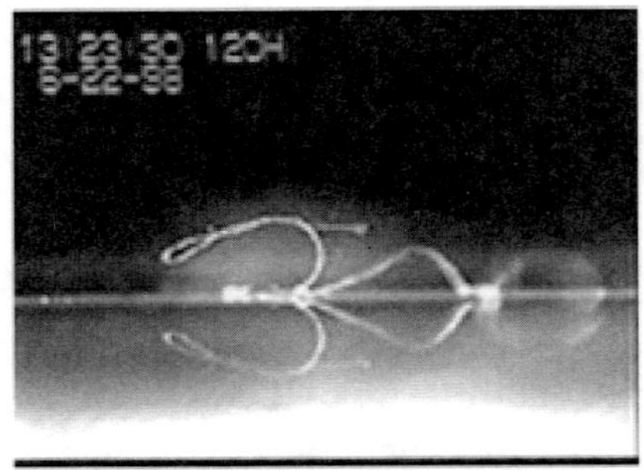

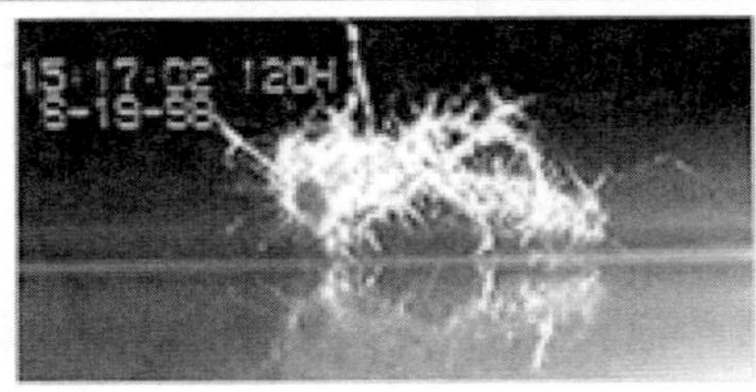

Fibers live on the floor of the growth chamber. They stand on it, roll over it, walk over it, drag things over it, and undergo self-assembly on it. No, the lower image shown above is not an insect out of focus, it's a late stage macrofiber perched on the glass surface where it was produced. A whole new world was opened by discovering the overlooked but obvious: all the action is down on the ground. It is somewhat frustrating to discover a whole new world when you are sixty one years old. NHM would have hoped that at the end of his career he could have drawn together everything into a nice package, a kind of what I have been able to do conclusion. But, under the circumstances, it didn't look as if all the issues now raised by his own efforts could be resolved by his own efforts. And so the question arose, what shall NHM do for his last research project? What is it he should tell the National Institutes of Health that he wants to do with his final research grant?

Before NHM wrote that last grant application he asked himself, what do you think the most significant unanswered question is about macrofibers and is it reasonable to try to answer it in the time remaining? He realized that the application would have to be based on the new findings. He thought it best not to mention that it would be his last application. And he knew that he would have to convince the reviewers that the proposed work could be completed during the four-year grant period. No one would miss the fact that NHM was 62 years old at the time the application was submitted so it couldn't be an application appropriate for someone at the start of a research career even though his new findings were indeed the beginning of a new world in macrofiberology. Here is what the proposal came to include:

i. a. the use of wire drag methods to measure forces associated with twist and supercoiling and to study fiber dynamics
   b. the development and application of mechanical instruments to measure forces directly and to manipulate fibers

ii. the characterization of macrofibers pivoting during growth

iii. the analysis of the mechanism by which macrofibers walk over surfaces

iv. the demonstration and measurement of dragging powered by a twist/writhe machine

v. the quantitative measurement of macrofiber helix hand reversal

vi. the following engineering and modeling projects:
   a. obtaining force measurements
   b. constructing a dual-image microscope
   c. the development of a half-fiber model of macrofiber dynamics
   d. a consideration of the restraint imposed by the floor
   e. developing a theory of walking using twist and writhe forces
   f. the use of twist and writhe to power a machine

It was indeed an ambitious proposal but NHM really thought that there was a good chance of getting it all done. As it turned out, he didn't quite make it, but it was close. Nothing at all was accomplished on project "v". Most of the rest was achieved. And without saying so in the proposal NHM was able to measure the supercoiling force in macrofibers, to his knowledge the first such measurement in any supercoiling system. For NHM to get the funds NIH first had to carry out its review, debate the merits and weaknesses of the proposal and in the end assign a priority score to it. To show you what that involves a copy of what NIH sent back to NHM is provided here. It is the kind of document that wrenches the heart of researchers who devote almost all of their lives to science and have no other way to get the money needed to support their research addiction. The Guardian of the core hated these documents right from the beginning. He saw them as violations of the inner sanctum of the scientist. Not only were they humiliating and dogmatic, they were ruthless in their design. He believed they were subject to scientific politics as well as the constraints of existing dogma. The shame of it was, he said again and again, that intellectual pursuits and the ability to go beyond the known had to be linked to money. And above all he suspected that creative work fell outside of the protected domain of fields and institutions that garnered the major part of the resources that were available. It was as if, he said, the best among them was made to dress as a jester and made the laughing stock. Guardian's partners appreciated his view, indeed they endorsed it, but they understood the force of addiction and how it was linked to survival. Addicts they said will suffer for what they need. When the need is strong enough, they will even die trying to satisfy it.

Unlike college admission letters the size of the envelope sent by NIH gave no clue as to whether its message was favorable or not. Ten pages, more or less, had to be read by the losers as well as the winners. Here is the document that NHM received.

SUMMARY STATEMENT

Application Number: 2 R01 RR07912-05A1

ZR01 MBC-1 (2)
Review Group: CENTER FOR SCIENTIFIC REVIEW SEP
Meeting Dates: IRG: FEB/MARCH 1999 COUNCIL: MAY 1999
MMR13
Requested Start Date: 06/01/1999

MENDELSON, NEIL H. PHD
UNIVERSITY OF ARIZONA
MOLECULAR & CELLULAR BIO DEPT
PO BOX 210106
ROOM 354
TUCSON, AZ 85721-0105

Project Title: COMPLEX PROCESSES IN MULTICELLULAR BACTERIA

IRG Action: Priority Score: 177 Percentile: 17.7#
Human Subjects: 10-NO HUMAN SUBJECTS INVOLVED
Animal Subjects: 10-NO LIVE VERTEBRATE ANIMALS INVOLVED

GENDER, MINORITY, & CLINICAL TRIAL CODES NOT ASSIGNED

PROJECT YEAR	DIRECT COSTS REQUESTED	DIRECT COSTS RECOMMENDED	ESTIMATED TOTAL COST
05A1	134,957	134,957	201,718
06	102,935	102,915	153,825
07	107,032	107,032	159,979
08	111,313	111,313	166,377
TOTAL	456,217	456,217	681,899

NOTE TO APPLICANT FOLLOWS SUMMARY STATEMENT.

RESUME AND SUMMARY OF DISCUSSION: This is a complex proposal from Dr. Mendelson to define the mechanisms of bacterial macrofiber dynamics. A goal will be to establish a relationship between individual cell growth and the morphogenesis of the macrofibers. The approaches taken will depend on both experiment and development of theory. The reviewer is satisfied the experiments can be completed as described. It is not clear however, how the results will be applied. There is no indication in the proposal how macrofiber properties can be used to define growth processes of individual cells. The proposal lacked a discussion of limitations of methods, and of potential errors and pitfalls. In numerous instances statements were made in the proposal, but with no effort to justify the statements. Neil Mendelson has virtually single-handedly created the subdiscipline of bacterial macrofibers. He has had a career characterized by creativity and has published in high impact journals. His record suggests that new and important information will emerge from the proposed studies. It is recommended the proposal be approved with enthusiam.

This is an innovative proposal from an experienced investigator who has enriched biology by recruiting mathematicians and physicists to explain bacterial growth patterns. He is unique in this regard. His work has been imaginative and the proposal continues this innovative spirit. My enthusiasm is

Date Released: 05/04/1999 Date Pinted: 05/07/1999
MBC – 1 R01RR07912-05A1
FEBRUARY 1999 MENDELSON, NEIL

fairly high but tempered by the following: 1. There were parts that were incomprehensible. 2. If one is interested in helical structures and chirality, then why not study DNA or protein superhelices where the structure is critical to the biology? The macrofibers happen to have writhe but that writhe can be right-handed or left-handed. The chirality of DNA and protein superhelices is essential to their function. I do not agree that "there is a universal stratagem concerning the physics and mechanism of living systems." 3. There was no attempt to gain molecular explanation of the forces involved – how they relate to bacterial physiology. He did not use mutants and inhibitors of this well-studied organism.

The research focused on matters such as measurement of forces associated with macrofibral twisting and supercoiling and kinetics, and the pivoting and walking of macrofibers, for example, represents sophisticated lines of imagination and inquiry. The need for development of protocols and instrumentation quite new for study of biological problems is made both apparent and reasonable and, in addition, likely to be successful.

At some point in time someone must clarify the chemical-biochemical aspects of these fibers, their biosynthesis and their localization in cell macromolecular arrays, if there be such latter. But that is not the focus of this application nor should it be -- fundamental aspects of behavior need to be more fully appreciated and more carefully definable.

Neil Mendelson should be supported on this proposal even if this reviewer would rather see him pursue a genetic approach. He is innovative and is totally committed to this approach and to the types of measurements proposed. His papers on the subject are thorough and of high quality. I trust that new insights into the hierarchical relationship between individual bacterial cell growth and the morphogenesis of multicellular bacterial macrofibers will come forth through his efforts.

DESCRIPTION: The ultimate objectives of the proposed research are: i. to understand the hierarchical relationship between individual bacterial cell growth and the morphogenesis of multicellular bacterial macrofibers, and ii. to characterize the fiber state. Macrofibers are helical structures that twist and writhe as they grow and thus provide a unique opportunity to study fundamental aspects of twist and supercoiling. The heath relatedness of these studies pertain to issues dealing with the behavior of DNA and chromosomes, to the use of twist and supercoiling as potential molecular motors in cells, and to the properties of multicellular bacterial states. There are seven specific aims all related to new discoveries: i. wire drag, and ii. Micromanipulator approaches will be used to study half-fiber behavior and to measure forces associated with twist, supercoiling and motions associated with fiber self-assembly. Newly discovered fiber motions: iii. pivoting, and iv. Walking will be characterized and their relationship to twist and writhe will be sought. v. The dragging of structures by macrofibers undergoing a twist to writhe conversion will be examined and the work accomplished measured. The power generated by fibers will be estimated and its relationship to fiber growth, and ultimately individual cell growth will be determined. vi. Macrofiber helix hand reversal will be studied in terms of structural constraints and the relationship of helix hand in pivoting, walking and dragging. vi. Two engineering and four modeling projects will be undertaken. Instruments that can measure forces associated with fiber motions will be designed and built. A dual-image microscope will be constructed for use in studying fiber motions that involve interactions with the floor of the growth chamber. Theory will focus on: fiber models that examine how the two halves of a fiber influence one another, examination of the significance of constraint on fiber morphogenesis caused by interactions with the floor of the growth chamber, the development of theory of walking caused by twist and writhe, and the properties of a twist/writhe machine that can do work. All seven projects included in this proposal represent the continuation of lines of investigation underway currently in the laboratory using experimental approaches developed here. They build upon foundations established by our long term study of the bacterial macrofiber system and seek to consolidate what is known into a comprehensive overview of how the system works.

The planned work is obviously highly innovative.

Investigators: Neil Mendelson is a Professor of Cellular and Molecular Biology at the University of Arizona in Tucson. He is well-known in Bacillus circles and has published several papers in impact journals. His work on Bacillus growth, especially relating to macrofibers, make him a world leader. He lists, however, only 18 papers in a >20 year period. John Thwaites (Cambridge) is a Professor of Engineering who has provided a strong collaboration with Mendelson. Thwaites is an acknowledged leader in physics and mechanical properties of polymers. The talents of Mendelson & Thwaites combine to form a strong research team.

Environment: There is no concern about facilities, as most of the equipment needed to complete the work as outlined is already in place. The facilities to be used by Professor Thwaites were not described, but it is assumed laboratory space, computers, etc., will not be rate-limiting. The research atmosphere at UA is superb.

CRITIQUE 1:

Significance: The research will be conducted primarily by the PI and his close collaborator, John Thwaites of Cambridge University. Further collaboration has been established with a physicist and a mathematician. Collectively, considerable theoretical and experimental talent will pursue the bases for macrofiber behavior. Results from the research may lead to new ways of thinking about cell-cell interactions, about physical properties of cell walls and about the role of individual cell growth in the physiology of a multicellular macrofiber.

Approach: The strengths of the proposal include: The persistence of the PI to almost single-handedly develop the field of bacterial threads and macrofibers. Mendelson has shown high levels of creativity in several publications in high impact journals. He understands Bacillus physiology and at the same time has applied polymer physics to the characterization of the macrofibers.

The fact that so little is known about bacterial macrofibers requires a serious study of their properties. The proposal is an extension of studies began over 20 years ago demonstrating restricted ranges in their forms. Would not some knowledge of the chemistry of their cell walls be useful here?

The proposal was difficult to read and analyze. Statements were made without any obvious explanation. For example, in the Progress (p27), it was stated "A mathematical pattern formation model was able to generate a similar pattern using inputs primarily based on the physics of convection suggesting minimum requirements for regulation by the cells". The reviewer was unable to determine what was meant and even if it was relevant to the proposed studies.

Throughout the proposal it was alluded that studies on macrofibers will yield information on individual cell growth. The reviewer was unable to understand how this relationship would be established, as specifics were not mentioned in the proposal.

The macrofibers may have novel properties previously unknown to polymer scientists, as the properties already identified depend on both growth and physical attributes of the fibers.

The system of macrofibers is amenable to genetic manipulation and to chemical and physical characterization.

The strong background of the PI, more of which will be mentioned below. The weaknesses of the proposal can be summarized around the following: #9.

The PI has known for years that filament properties (i.e. RH to LH to RH) can be controlled by various agents. Yet, there are no published reports on wall chemistry. Is RH controlled by chemistry distinct from HL? There are no results describing composition, crosslinking or chain length of walls from Bacillus macrofibers. The reviewer agrees that it is not absolutely necessary to know wall chemistry in order to perform the experiments described in the proposal, but argues that all of the properties of macrofibers ultimately rest on wall structure. On p16, the PI stated mutants had been generated.

The PI does not consider the use of metabolic inhibitors on any of the macrofiber properties. B. subtilis elongates its side walls by random intercalation of new wall into old wall. Side walls constitute about 70% of the surface area, whereas septa and poles account for the remainder. There are antibiotics which seem to prevent septa assembly and some which prevent both septal and side wall assembly. Why not add an antibiotic to the macrofiber as it undergoes movement. It would also be interesting to employ chloramphenicol, which prevents wall extension, but promotes wall thickening. It would be interesting to add an uncoupling agent during measurement of a particular motion. How and where is metabolic energy applied to the macrofiber? In addition, at least one of the FJ series mutants is temperature sensitive. By going from one state to another following temperature shifts, it may be possible to determine the role of autolysin in macrofiber genesis. #9

Innovation: Mendelson has virtually owned the filament-macrofiber field of Bacillus this past 15-20 years. He now proposes to extend his observations to develop the biophysics of macrofibers to a new level, and to apply new methods for the characterization of macrofibers. Mendelson and his collaborators may be the only people in the world capable of defining macrofiber biophysicist.

OVERALL EVALUATION: This is a complex proposal to define the mechanisms of bacterial macrofiber dynamics. A goal will be to establish a relationship between individual cell growth and the morphogenesis of the macrofibers. The approaches taken will depend on both experiment and development of theory. The reviewer is satisfied the experiments can be completed as described. It is not clear however, how the results will be applied. There is no indication in the proposal how macrofiber properties can be used to define growth processes of individual cells. The proposal lacked a discussion of limitations of methods, and of potential errors and pitfalls. In numerous instances statements were made in the proposal, but with no effort to justify the statements. Neil Mendelson has virtually single-handedly created the subdiscipline of bacterial macrofibers. He has had a career characterized by creativity and has published in high impact journals. His record suggests that new and important information will emerge from the proposed studies. It is recommended the proposal be approved with enthusiasm.

CRITIQUE 2:

Significance: The area of study is of wide interest as bacteria growth in aggregates is their predominant form in nature. The questions addressed importance also in terms of patterns and growth. The studies by Mendelson have provided a model of mathematical and physical approaches to biology. Indeed, the productive multidisciplinary nature is the greatest strength of the application.

OVERALL EVALUATION: This is an innovative proposal from an experienced investigator who has enriched biology by recruiting mathematicians and physicists to explain bacterial growth patterns. He is unique in this regard. His work has been imaginative and the proposal continues this innovative spirit. My enthusiasm is fairly high but tempered by the following: 1. There were parts that were incomprehensible. 2. If one is interested in helical structures and chirality, then why not study DNA or protein superhelices where the structure is critical to the biology? The macrofibers happen to have writhe but that writhe can be right-handed or left-handed. The chirality of DNA and protein superhelices is essential to their function. I do not agree that "there is a universal stratagem concerning the physics and mechanism of living systems." 3. There was no attempt to gain a molecular explanation of the forces involved – how they relate to bacterial physiology. He did not use mutants and inhibitors of this well-studied organism.

CRITIQUE 3:

Significance: Cell growth (division) is the most fundamental of all biological processes; the research proposed aims to elucidate, seemingly primarily in physical terms, a range of traits of macrofibrial changes in form and movement. These may very well relate to crucial and fundamental aspects of cell division that are beyond the scope of other studies of cell division because they are beyond the perceptions available to other investigators. Although seemingly not envisioned by the PI, an understanding of the events and phenomena outlined may also provide new and fundamental insights into the ability of many pro- and eukaryotes to migrate on solid surfaces in the apparent absence of visualizable locomotor organelles. The human tooth surface is but one example of a habitat in which bacteria of such type thrive and may be involved in diseases.

Approaches: To one who is not a physical chemist but who appreciates biophysical approaches to analyses of biological phenomena, the "intellect" of the analytical procedures outlined seems sound. The fact that the "team" of researchers includes a physicist and mathematician is not only laudable, but essential. A great deal of my belief that the outcomes of the research will be of cardinal importance reflects the PI's past performance(s) in bringing the recognition of these basic problems and their potential significance from a phenomenological level to scientifically tractable ones. The PI has been alone is recognizing the developing these studies.

Innovation: As noted above, the subject matter of the research area is not only extraordinarily novel, but this to my knowledge is the only laboratory in the world equipped, either intellectually or physically, to tackle these significant questions. In the most general of terms, one could argue that since "macro" versions of the approaches to be employed are used, directly or indirectly, in different engineering programs, that the approaches are not novel ones. However, the miniturization of the analytical approaches and the application of these to what have heretofore been essentially unrecognized macromolecular "dynamics" IS novel!

Investigator: As suggested above, the PI has excelled in bringing our understanding or recognition of unanticipated traits to the point that they are now. He has recognized the need to not only collaborate with, but to INVOLVE, those with more physical and mathematical perspectives in bold attempts to elucidate the behaviors described (and which are made more dramatic by actually viewing them at "his" website). The clever adaptations of analytical procedures represent a real "tour de force".

I emphasize my belief that this PI and his collaborators are THE ONLY ones who sufficiently understand the subject matter and have the wherewithal to made productive contributions leading to our increased perception.

Environment: As is surely clear from the comments above, this PI and his collaborators, and their laboratories, are THE people and places where we have any right to expect broadened understanding and deeper comprehension of the nature of macrofibrial movements and the biological significance of these. Had the PI and colleagues not brought to field to the point it is now, one might well be dubious of their ability to understand and tackle these problems.

Approach: The approaches are imaginative and informative. I was disappointed, though, that he did not take advantage of the ease of study of B. subtilis to relate the macrofibers to intracellular events.

Innovation: The proposal is highly innovative.

Investigator: Mendelson is not the best person to do the work; he is the only person. He has great collaborators, particularly Taber.

Environment: This is ideal. Mendelson has set up rich collaborations in Arizona.

OVERALL EVALUATION: The research focused on matters such as measurement of forces associated with macrofibral twisting and supercoiling and kinetics, and the pivoting and walking of macrofibers, for example, represents sophisticated lines of imagination and inquiry. The need for development of protocols and instrumentation quite new for study of biological problems is made both apparent and reasonable and, in addition, likely to be successful.

At some point in time someone must clarify the chemical-biochemical aspects of these fibers, their biosynthesis and their localization in cell macromolecular arrays, if there be such latter. But that is not the focus of this application nor should it be – fundamental aspects of behavior need to be more fully appreciated and more carefully definable.

CRITIQUE 4:

Significance: After the discovery of helical forms in 1975, the PI singularly has been involved in extensive characterizations of these unusual filamentous structures that has led him to propose models on cell wall assembly which follows a path that accounts for the twisting motions. He wishes to establish how two halves of a fiber influence one another, how bacterial fibers can walk on a solid substrate and the ability of macrofibers to move objects like a bacterial ball. Basically, according to Mendelson "in macrofibers force is transduced via geometry and motion into twisting and writhing and the physics of these processes govern how the forces operate. – The key to using the potential twist and writhe appears to be the impediment to motions, the same mechanisms responsible for the cylinder to helix deformation and self-assembly in macrofiber".

Approach: The entire proposal is a biophysical approach of bacterial macrofibers rather than a molecular genetic approach. It relies on just two equations presented on p29 and also discussed in his review article (in Bacteria as Multicellular Organisms, 1997). The reference for the equation is missing, certain basic definitions of twist versus writhe, the interaction potential between the fibers have not been elaborated and other parameters have not been properly defined for a non-physics reader. For example, "the forces responsible for folding and winding up must be related to the twist to writhe conversion" the question is how? The PI wanted to emphasize in his response to the criticism raised previously that now he is presenting a proposal as a geneticist yet the approach remained the same.

It is difficult to critically evaluate the seven projects without additional experimental details or cartoons that describe how the micromanipulators will be applied, optical traps, Young's modulus and how one generates fibers with different twist as stated by the PI through the use known protocols (no reference).

The experimental approach has been used in part by the PI and in the included preprints and I am convinced will generate a lot of data. Also with his engineering collaborator Thwaites, they designed a most sensitive transducer that allowed them to make force measurements and they are planning to design even more sensitive transducers. After completing the various projects and accumulating the data on the forces associated with twist, supercoiling, fiber walking and dragging of macrofiber-ball structures, he plans to make this information available to his engineering, mathematics and physics collaborators for consideration in their model building efforts.

The PI is convinced that the physics of twisting and writhing on a surface will provide insights into the mechanism by which cell growth results in motions. He may be correct and should be encouraged to pursue these studies since he has been productive and innovative in this field and is likely to demonstrate that "macrofibers show how growth itself can set into play a self-assembly process as a result of physical constraints rather than genetic programs". He may be correct that "twist and writhe are the source of energy used in fiber morphogenesis and in fiber motions". Yet genetic approaches have always been essential to test the validity of conclusions derived from in vitro or in this case biophysical experiments. The projects that were eliminated from this proposal were just summarized in the progress report on cell viability, organized cell swimming in colonies, mutant studies and reporter gene expression in macrofibers were the most interesting to me and should be continued. They hold promise if the fusions are better characterized. This reviewer was wondering about the statement "several new constructs in which both map location of the report gene as well as the physiology by which the host gene is normally regulated are known" – who are these genes? Moreover, the preprint on organized cell swimming contained important observations on the rates of swimming, frequencies of switching from counterclockwise to clockwise direction and that sessile cells in dry regions do not lose the ability to swim (their flagella). These observations should be pursued with a mutant study using the many available fla and che mutants that have been characterized in B. subtilis. Also, work on peptidoglycan and bacterial cell walls should continue.

Innovation: This proposal, and Dr. Mendelson's work in general is obviously highly innovative.

Investigator: Dr. Mendelson is the world's leader in this area of research.

Environment: I am also impressed that some of the work is done by talented undergraduate students who are either math, physics or molecular biology/music majors who were involved in the single – wire impediment studies or in the side and dual – view films. His main collaborators like Drs. Tabor and Goldstein as well as Dr. Thwaites have been important to him as well as he inspiring them to continue in his area. Also, other collaborators like S. Mann and H. Bennich are important in making this area of investigation highly relevant.

OVERALL EVALULATION: Neil Mendelson should be supported on this proposal even if this reviewer would rather see him pursue a genetic approach. He is innovative and is totally committed to this approach and to the types of measurements proposed. His papers on the subject are thorough and of high quality. I trust that new insights into the hierarchical relationship between individual bacterial cell growth and the morphogenesis of multicellular bacterial macrofibers will come forth through his efforts.

ad hoc or special section application percentiled against “total CSR” base

NOTICE: The NIH has modified its policy regarding the receipt of amended applications. Detailed information can be found by accessing the following URL address:
http://www.nih.gov/grants/policy/amendedappls.htm

++Beginning with the round of reviews for October 1998 Councils, NIH peer review groups were asked to recalibrate their scoring using 3.0 as the target median score and to spread scores over a wider range of the priority score scale. (This measure is being taken for two reasons: 1) inconsistent scoring practices among different review groups, and 2) in many review groups, scores have become so compressed that the ability to discriminate among applications has been compromised.) As a result of this recalibration, priority scores for applications reviewed this review cycle may not be comparable to those received prior to the October 1998 Council review cycle. Percentiles for applications reviewed this round are based on scores assigned the previous two rounds and this round.

Observer went right to the bottom line when he saw this summary statement. The seventeen and seven tenths percent priority score and the recommendation that all the funds requested should remain in the budget were the meat. He knew the funding cutoff was then in the range of the top twenty to twenty five percent of all proposals deemed worthy of support therefore the grant was very likely to be funded. Creator knew at the time the application was written that only one out of four or five applicants that should be supported would be. The odds were against us then. It didn’t help matters either that what our proposal was about had nothing to do with the mainstream focus of modern biology. That was not our game. We were out there on our own as always, dependent upon just one agency for our support, and without any political capital to help our cause. All we had to go on was Creator and our discoveries. We thought our story was a compelling one and pitched our proposal on the chance that physics would be the next frontier in biology. We must have made the case well enough. Look at what they said about us. NHM himself was amazed at the enthusiasm, the praise, and the appreciation of his accomplishments. That is what will remain with him from this exercise. It tinted his eyeglasses a rose color. He interpreted criticisms suggesting that he should focus his energies elsewhere as recognition and appreciation of the work he did in those "elsewhere" things early in his career. Nothing could take the glow off comments like: "It is recommended that the proposal be approved with enthusiasm", "His work on Bacillus growth, especially related to macrofibers, makes him a world leader.", "Mendelson is not the best person to do the work, he is the only person.", "...the subject matter of the research area is not only extraordinarily novel, but to my knowledge is the only laboratory in the world equipped, either intellectually or physically, to tackle these significant questions." As you will see later in this book this is not the kind of confidence people had in NHM for most of his life, nor is it the kind of recognition he received for what he had attempted. No, he was not that fortunate. But as he suggested in his letter to the student Paul, he did not become discouraged by what others told him he could not do! NHM was, in a way, immune to their criticisms. "As you know he is immune to social pressure." a comment made about him in a talk given by one of the MCB younger professors wasn't far from the mark. "You are a survivor, Neil." A comment made by Marty Slater the NIH Scientific Review Administrator who was responsible for his grant application reviews. In order to survive NHM had to disregard those who said, "you're no damn good", or even, "you're just not good enough".

9.What hurt the most

To NHM what people had to say was harmless so long as he could control the amount of time they took from his productive work. He could accept being suckered, such as the time the decision package was stolen. He could accept suffering fools; which is the story of his career at Arizona. But, the thing that hurt him the most was the dismal financial situation that plagued his family once he arrived in Arizona.

When NHM left the Army in 1966 his net worth was $5,000.00. Accumulated vacation leave not taken covered the cost of his family's voyage to England, and his NSF postdoctoral fellowship of $7,500.00 provided for all their needs in London, including the purchase of a car before they returned to the United States. NHM's income while an Assistant Professor at the University of Maryland, Baltimore County, enabled them to purchase a beautiful small house with no down payment under a veterans benefit program. They easily met all their expenses on his salary. When they got to Tucson their net worth after making the down payment on a modest house and using the three years worth of retirement funds accumulated in Maryland for their move was $800.00. There were no luxuries. They literally had to do everything, house, car, clothing, by themselves. When their net assets sank to about $300.00 Joan gave up all aspirations of becoming a scientist and quickly entered law school. NHM searched for a second job. He obtained a real estate license. But Joan insisted that he hold off just a little longer. Her suggestion worked. He was lucky enough to get the Research Career Development Award from NIH. Even that, however, ended in somewhat of a financial disaster. NHM had sufficient money to take his family to Paris for the fifth and final year of his award, having found a renter for his house in Tucson. His travel to Paris could be taken from his NIH research grant. When they returned to Tucson there was a little surprise waiting for them at the University. Owing to an accounting screw-up, the University had overpaid him when he was away without detecting their error. On a monthly basis it was a small amount that neither he nor his wife could detect. Two things made detection difficult. First NHM's salary was always increased for a three-month period during summers when he was able to take salary from his research grant rather than either University funds or the Career Award funds. On top of that there was always a currency fluctuation in converting dollars to French francs, so what was remotely deposited in their bank in Paris was a different amount each month. Those factors were of no interest to the University Administration. All they cared about was getting their money back, which NHM simply did not have to return.

The Executive Vice –President and Provost called NHM to his office. He refused to apply any funds that the University had obtained from the overhead of NHM's grants to meet this debt. He insisted that either NHM pay or they would take it out of his salary. NHM told him he would look into the legal issues and let him know what he planned to do. The Provost immediately forwarded the matter to the University Attorneys. Joan, then an attorney, and NHM decided that they would have to "bite the bullet" and pay off the debt. The University of Arizona had made them debtors. NHM agreed to the repayment plan as detailed in the following letter to him from the Administration:

THE UNIVERSITY OF ARIZONA

TUCSON, ARIZONA 85721

EXECUTIVE VICE PRESIDENT

February 6, 1978

Dr. Neil Mendelson
Department of Microbiology
Pharmacy-Microbiology 217
Campus

Dear Neil:

This is the letter I promised to send you during our talk of last Thursday. We agreed that I would see that your W-2 Form reflected your proper earnings rather than what was paid you in error and that the overpayment which you actually received would be repaid over the next few months. The situation is this:

	Actual Earnings in 1977	Paid in 1977
Gross	$ 27,489.00	$32,694.15
Federal Tax	4,909.96	6,758.38
FICA	965.25	965.25
State Tax	490.98	675.84
Retirement	1,924.20	2,288.61
Health Insurance	231.12	231.12
Miscellaneous	112.20	112.20
Total	$ 18,855.29	$21,662.75

The net overpayment in 1977 was $2,807.46.

There was also an incorrect check issued to you on January 15, 1978.

	Correct Amount	Check Received
Gross	$ 1,156.70	$1,735.10
Federal Tax	205.81	411.21
FICA	69.98	104.97
State Tax	20.58	41.12
Retirement	80.97	121.46
Health Insurance	19.57	19.57
Miscellaneous	13.04	13.04
Total	$ 746.75	$1,023.73

The new overpayment due to this incorrect check was $276.98 and the total overpayment amounts to $2,807.46 + $276.98 = $3,084.44.

This sum could be repaid to the University in one lump sum if you should choose to do so, but it is my understanding that you are not in a position to choose this option.

As we agreed in our discussion of Thursday, then, the overpayment of $3,084.44 will not be reported to the I.R.S. or to the state as salary. For this reason, I will have to ask you to sign a promissory note for $3,084.44 in favor of the University. This amount will be repaid over a period of 7 1/2 months (assuming you work during the summer on a grant or contract) in installments of $205.62 per pay period. This will require 15 payments at this rate.

I propose that, beginning with the check for February 15, 1978, you endorse your check over to the University and receive one in exchange for $746.72 - $205.62 = $541.13. You can, of course, if you prefer, receive the check for $746.75 and at the same time make out a personal check to the University for $205.62.

I trust that you will find this a reasonable and equitable procedure. It is the fairest arrangement I can work out.

Needless to say, the Business Office (and I, also) regret very much that a mistake was made on your checks and that it has caused you both inconvenience and some annoyance. I am assured that every measure is being taken to ensure that mistakes of this kind will not be repeated in the future. Your cooperation, forbearance, and understanding are very much appreciated.

With the best personal regards,

Sincerely,

Al Weaver

A. B. Weaver

ABW:nh

P.S. I, of course, realize that your salary will change during the summer & next fall. The deduction will continue to be $205.62 until the $3084.44 is repaid.

ABWeaver

Weaver's humiliating news couldn't have come at a worse time in NHM's career. On February first ,1978 NHM's twin daughters turned thirteen. His wife was still new to the practice of law, which she began in 1975. NHM had published two of his most important papers, one in 1976 the other in 1978. And he was virtually broke. There were twenty six years remaining to dig out of that hole although at the time NHM didn't know that. Today his back is sore from the digging. Well the digging and perhaps the playing because in the interim he found that the kind of music he loved to play, Bach, Handel, Faure, Mozart, Brahms, Vivaldi, was much sought after in many churches in Tucson. They were willing to hire NHM as part of their performances. It never brought him a lot of money but it did bring a lot of joy.

In 1973 when NHM received the Research Career Development Award from NIH and went to see Dr. A. B. Weaver about the tax consequences of it, it was an occasion for congratulations. Isn't it great you got the first one ever awarded to the University, Dr. Weaver said. Who would have thought then that five years later it would come to a three thousand dollar debt, and Dr. Weaver would be responsible for collecting it? At the time it was all good PR for the University. That's what they saw in it. Someone in the Administration must have put a press release together although NHM never saw one. The student newspaper published a piece about the RCDA and so did the Tucson newspapers. People wrote offering congratulations and expressing the hope that NHM would cure cancer and the like. He actually met one of them when they were both sitting waiting to be called at a clinic in the University Hospital. When NHM's name was called, a man came up and introduced himself. He seemed troubled that NHM too could be sick although he himself was a retired physician. On the way home driving through the desert heat in his old decrepit Saab NHM was almost knocked out of his groove.

The radio was playing, one of the few things that still functioned in his old car. On came, "Cele Peterson now announces the Star of the Day, Dr. Neil Mendelson" followed by a blurb about his being the first to win a Research Career Development Award from the National Institutes of Health (perhaps she thought it was the first one they ever gave, not just the first at the UA). Peterson, the owner of a local clothing store and a philanthropist, went on about NHM's cancer-related research. NHM nearly drove into a tree. The next day an award certificate arrived in the mail. NHM had lived through twenty seven such announcements on the radio the day before. And nothing bad happened as a result of it afterwards either. A few months ago NHM met Cele Peterson for the first time. He and his wife were invited to go with friends to a concert of the Tucson Chopin Society a group they had never heard of before. Seated at a table before the concert began Joan said, "I believe that is Cele Peterson over there." How would I know NHM responded? Is she still alive? Oh yes Joan said, and her dress shop is still in business. When there was a break in the program NHM went over and introduced himself to her and told her that he had been one of her "Stars" thirty one years ago. She was thrilled that someone could remember that! Oh yes, he told her he still had the certificate. What do you do now she asked? I am a farmer, NHM replied. Everyone seated at her table turned to look at him in silence as she did. NHM continued, yes we grow lavender flowers on our farm in Oregon. Lavender, Peterson exploded, I love lavender. What do you do with the flowers she asked? Several things NHM said, our main goal is to develop new varieties that have optimal health properties such as calming agitated patients, improving burn healing and things like that. Oh she said, it's wonderful, would you come to the studio? I'll give you another Star of the Day award she said. Really NHM asked? Yes, please come. Sure, he responded. Her protocol had changed. This time the "Stars" sit down for a little discussion with her, then she formulates an introduction for broadcast that is followed by a short statement made by the Star. The day before it was to be broadcast NHM got an email notification, and the day afterwards the certificate arrived. Here it is:

"To Give of Oneself Is the Greatest of Gifts!"

We Salute Our

Star of the Day

Neil H. Mendelson

In Recognition

Your research into therapeutic
Effects of Lavendar

KSAZ 580AM salutes you on

April 14, 2004

Congratulations and Thank You

Sincerely,

KSAZ AM 580

NHM always knew he was at least a two star person!

On June thirtieth 2004 NHM planned to retire from the University of Arizona after thirty five years of service. The University did not let its employees forget how long they had been there. Every five years professors, administrators, janitors, gardeners, secretaries, whomever, were invited to a luncheon at which the President and his underlings gave praise, thanks, and gifts as well as food. Given the size of the institution hundreds of people were in attendance at these celebrations. Those being honored for five, ten, fifteen, in some cases as many as forty five years of service packed together in a ballroom to hear how well they had done and how much they were appreciated. In this most recent gathering NHM feasted on chicken followed by chemical desert. It should have come in a blue bottle. NHM examined it carefully. A professor of Nutritional Sciences seated by chance at the same table refused to assume any responsibility for its composition. I guess you could say this was a tasteless affair, NHM thought. As gift time approached NHM couldn't help thinking of all the treasure bestowed upon him at previous occasions. There was the letter opener at ten years that was a lethal weapon no one in his or her right mind would keep in the office lest a disgruntled student would be tempted to use it. At twenty five years he got a small wooden box with notepaper in it. They missed the boat he thought it should have been shaped like a coffin. The thirty-year prize was unique, a small framed etching of the Old Main building where the University got its start on land donated by gamblers prior to statehood. This was clearly meant to refresh the memory of those who might have forgotten where they had slaved. Somehow that might not have seemed enough. A new surprise was added.

Each person had his picture taken with President Likins (third from the left in the blue shirt) along with Wilbur and Wilma Wildcat. NHM felt a little out of place up there with them having never been to a UA basket-ball or football game, but his grandsons saw it differently. He was definitely now a hero in their eyes. At his thirty fifth year ceremony NHM wondered how they could possibly outdo his last one. When his name was called he walked bravely across the ballroom thinking this is quite a distance for an old man to walk. This time Likins stood with

NHM plus his gift for the photograph. The gift was a coffee-table book of photos showing the Grand Canyon. The President had inscribed it with a marker pen. NHM was thrilled to get it, having only once seen the Canyon a few years after he arrived in Arizona. In those days his car was not really reliable enough to take on any long trips. Observer can recall only four journeys besides their going to Europe in the years when their children were pre-college age. They went to the Grand Canyon, to the Indian Pueblos in New Mexico, to San Diego, and to Rochester, New York where they lived in Frank Young's house during the summer that he went to Portugal. One wonders where the tradition for these service awards came from, who invented it, when and for whom it was intended. Is it one of those things that once started continues in perpetuity never to be reevaluated or thought of again? Who picks the gifts that are awarded? Has the twenty first century gotten to Arizona yet? Which calendar are they using?

When it became apparent that NHM's life span as a research scientist was approaching its limit, time and again he asked himself how he could have continued at it for as long as he did given all the impossible things that had to be endured in the workplace? Creator thinks he might know the answer now as a result of recent terrible disruptions that arose in NHM's laboratory. Ever since NHM first set foot into the research environment he found the laboratory to be the most ideal place to think, to challenge himself, and to discover how living things work. It was the only place he ever had been where one could shut out all trivial disruptions, and focus without interruption. There were no bosses in the laboratory, no distracting obligations, no hoodlums, none of life's spam, not even telemarketers! But to maintain the tranquility it was necessary to strictly enforce a set of rules including: the very careful selection of those allowed to join the research team, not allowing casual visitors into the laboratory, no music, food, socialization, and no noise. The laboratory was a place where you could focus your entire energy and to NHM that was such a rewarding thing that it kept him from giving up the profession in the face of its undesirable aspects. Unfortunately, toward the end of his career things indeed did begin to unravel. The difficulties in the laboratory arose when NHM lowered his standards just a notch and it came back to haunt him. Three errors and NHM couldn't wait for the inning to be over.

Creator argued that lowering his standards was not the cause of NHM's problem. Rather he saw it as a conflict of interests in which the needs of an experimentalist working at the cutting edge of science were pitted against the university administration's needs to protect laborers as a union would in the world at large. It's nothing new he insisted and he supported his position by citing numerous cases of talented and successful experimentalists who worked only alone, wherever they were, no technicians, no students, just the scientist by himself. In the long run you're better off working that way, Creator continued, although that's not the way things are usually done these days. NHM had to agree after having suffered three setbacks in a row. He was quite angry with Creator for not having spoken up earlier when the problem could have been avoided in the first place. Observer could see an internal conflict arising and wanted to avoid it at all costs. Let's discuss this in more detail he suggested. And all the parties agreed. Here is the way the argument went, all inside NHM's head of course.

NHM had his say to begin with. Look, I had a terrible pressure to get as much done as possible before my body and mind failed me, my research funds were depleted, and my competitors swallowed up my discoveries as their own. I didn't have the luxury of time to find a perfect match for what I needed, and in truth there may not have been one given that I had invented all the technology involved and made the discoveries upon which it was all based. "So what?" Creator said. Didn't you see the recent *New Yorker* cartoon in which a man sits behind a desk trying to hire someone for a large corporation? No I didn't NHM responded. Well, Creator continued the man is clearly exhausted and defeated. Across from him sits an applicant he has interviewed, drawn as the most perfect moron, smiling and totally out of touch with reality. The exasperated head hunter trying to salvage something says in the caption to the cartoon below, "What the hell, we can use an idiot." NHM retorted, yes but what happened afterwards, when the moron had to do his job? Exactly Guardian added, how can you get the job done when you need to under such circumstances? NHM didn't hire any idiots, Performer added. On the contrary he picked only very intelligent individuals. That is true Observer agreed but notice it takes more than intelligence alone to do experiments. Let's look at each case and see what went wrong, NHM suggested. Observer provided the details.

When research funds were temporarily lost NHM's long term research technician sought and found an alternate job at the University of Arizona. When funding resumed, a replacement had to be found as soon as possible. NHM visited the University's Personnel Office and learned of the new proceedures that had to be followed in order to hire anyone as a research associate. It was a cumbersome process requiring examination of a number of candidates before any one of them could be hired. NHM notified his colleagues at the University that he had a new position to be filled and evaluated the recommendations sent to him by them. There were no ideal candidates among the files he read but one of them was clearly more suitable than all the others.After inteviewing several candidates he chose the one he thought best and made an offer that was accepted. NHM made sure that he reserved the right to terminate this individual for cause if necessary and was assured that he would be able to do so. After several months evaluation on the job, NHM concluded that the new technician could not accomplish what had to be done in a timely and accurate manner. He then requested and received instruction for how to document the reasons for termination, and he executed everything precisely as instructed. The technician objected and claimed that termination was not possible for at least a year after having been hired. Impossible NHM thought, but it turned out to be true. The attempt to dismiss the technician was followed by a work slowdown. To get the project completed NHM had to take over all the experimentation. His NIH research grant funds to support technical help were simply wasted for the entire year. To elimate further disruptions NHM reassigned the technician to work only on line rather than in the laboratory . Creator and Performer were quite upset by this outcome, but Guardian calmly counseled that there was really nothing surprising about it.

The needs of an individual simply outweigh the needs of a research project. Science is one thing but people are a higher priority. Wait a minute Creator interrupted, it's not just the performance of the technician here that's at fault, it's also the incompetence of the University Personnel Office, the people who provided guidance to NHM, represented themselves as the authority in matters of hiring and firing, and gave instruction that was totally incorrect. They should be responsible for making up the loss to NHM's project and the National Institutes of Health whose funds were earmarked for a particular scientific purpose, not a humanitarian purpose. Yes but "get real" is all that Guardian said in response. Okay we've taken strike one, but remember a strike could be either the result of the perfection of the pitch or the mistake of the batter. Sometimes it's hard to assign fault. Either way that leaves two to go before we're out.

Often timing is the key to everything. The right person comes along at the right time and everything goes right. Failure two looked as if that would be the case but it didn't turn out in the end to be just that. No, far from it. Out of nowhere right here in Tucson Arizona there appeared a senior person educated in just the right stuff, lots of practical accomplishments, lots of access to state of the art computer programs that would be just ideal for NHM's research needs, and yes, even some knowledge of the bacterial macrofiber system. By hindsight, too good to be true Guardian added. Creator, Performer and Observer disagreed. They said that the facts very strongly suggested by foresight that it would be hard for anything but a home run to be the outcome. On top of that having strike one fresh in mind all four of NHM's voices thought they had been educated adequately to guard against any kind of repeat of the first failure. So, everything was done to show the Personnel Office how far ahead of all other applicants this one was. Somewhere deep down inside all of them nevertheless they knew that if for any unforseeable reason it didn't work out another year's worth of funds would be gone. How to deal with this really wasn't debated but Creator had a plan that was meant to curtail losses if need be. The idea was simple. Rather than to make a big deal of it Creator folded it into the overall work plan that would be assigned to the new individual in such a transparent way that none of the others ever noticed it. There were just three rules to follow:

1. Start with easy stuff then slowly transition to more difficult things. Begin with analysis of data already in hand.

2. After three months of data analysis start experimemtation by repeating those from which the initial data had been obtained. Assess skill level by comparing the new results to those found originally by NHM.

3. Only permit new experiments to be done when NHM has determined that skill levels are adequate.

The hope was that by the end of the first year the new associate would be fully up-to-speed: able to work independently to produce new data, capable of analyzing it, and to organize the findings for NHM's evaluation. As had always been the case in NHM's lab the initial experiments were worked out by him before being assigned to anyone else for further study. Then if something different turned up in the course of the other person's work NHM would confirm the findings by his own new experimentation. That's the way it worked for over 40 years. Graduate students, postdoctoral fellows and visiting professors were naturally afforded much greater leeway, but even they were guided to work within the framework of NHM's focus. That is what the NHM laboratory was about.

It was a great plan that should have worked, and it did in many ways but not totally. Little failures initially insignificant compounded and eventually the entire house of cards supported by them collapsed. Guardian was the first to worry and Observer sided with him on just about every point. For months they kept grumbling about the way in which data was kept and annotated. There just wasn't enough detail written down and what was recorded was often impossible to understand. Efforts to correct these deficiencies resulted in either marginal or no improvements. NHM bypassed the problem as best he could by going over each missing thing and taking notes on what the associate told him, but he too worried that memory would fade with time and findings would be lost because of it. Creator and Performer were very busy with their own things and paid little attention to this problem, until a form of the same thing emerged in their areas of responsibility.

New experiments being done using their ingenious plan to solve a key part of the macrofiber puzzle were not being done with the required degree of precision. They urged NHM to try to do something about it and he did. He explained to the associate exactly how things had to be done and why, and he was promised they would be, but they never were.With each passing week, month, season, frustration grew. All NHM's voices knew the time for terminating the associate was long past. They all knew that reassignment to another task was not an option any longer. After much soul-searching Performer said, look, I'll take over this project. Creator urged NHM to let him do it. Allow the NIH grant to terminate when it runs its course he added. Seek no additional funds. Use what is left in the budget to hire part-time students.With them finish up your days as an experimentalist and we're out of here. NHM took Creator's advice and squeezed out as much as he could. Yes he got the data he needed but to this day other obligations have prevented him from writing the final paper. The information it will contain has been presented however at a national meeting of the American Physical Society, so those whose work is most closely related to the measurements he obtained are aware of what he accomplished. There is a chance moreover that the information will find its way into the final chaper in a small book about macrofibers that NHM hopes to live long enough to write. If not, well, it will be there in the American Heritage Center archive that is collecting all of his work. Eventually someone will find it.

Strikes one and two show you that the conflicting needs of research and employment security constrain the rate of progress that can be made in science. No question about it Observer added. If you want stability do work that is "more but the same" of what is already known. Don't get too fancy. If you were born to be ahead of the pack, forget about using laborers to help your efforts. Run as fast as you can carrying it all by yourself. There must be another way, Creator chimed in. What have you in mind, Performer asked. Well, what if we used volunteers, Creator responded. Hmm, NHM thought, what if we did? That might work. Let's try it. A volunteer might ease our transition to green pastures NHM said. But then Creator got cold feet. This is not a great time to try something new, he cautioned. We'll have enought on our hands with end-game planning, divestment of laboratory equipment, and making the transition from one kind of life to another. Do we really want to start a new learning curve on top of all that concerning the integration of volunteers into our operations?

It's the last of the ninth inning, two out and two strikes on the batter. The pressure is really on. Here comes the pitch. Is it a strike or a home run? In NHM's case it's worse than just a strike. A strike would have ended the game and NHM's calendar would have been cleared for him to do whatever he needed to get done. Instead he was dragged into a time-consuming and dangerous situation. Events spiraled out of hand beyond just his laboratory or even the University of Arizona to involve the federal government. It was an incredible, almost unbelieveable incident that shows just how risky it is to even let a volunteer with strong quantitative skills and a burning desire to do something good set foot in the laboratory before you know all the details about him. In fact, as you will see, it is virtually impossible to learn what you need to know about a person before you have to make a decision about him. The rule of thumb seems to be that when there is something wrong with a person the more serious the problem, the harder it is for anyone to find out about it. The system works to keep that information private as if whatever was wrong with the person in the past will disappear by magic the next time he gets a job. Ordinary avenues of vetting a person do not penetrate this veil of privacy that keeps light from ever shinning on the dark past. Before revealing all the details we need first a little history about, "Megastrike" as we shall call him and how he got his foot in the door.

NHM was introduced to Megastrike many years ago by a person working his lab and said to be a microbiologist/mathematician who worked hard to bring scientific literacy to impoverished students in Mexico through a Church organization. Over the years NHM occasionally saw him walking on campus and assumed he was employed there in some capacity although he was not affiliated with any of the programs in which NHM participated. Once in a while they stopped to chat, usually about their mutual friend who had left Tucson with her husband to work in another state where he had gotten a good job. During the time that NHM was attempting to attract some highly skilled students to pursue an engineering project in the lab he happened to meet Megastrike on campus and asked him if he by chance knew of any students who might be interested. Not at the moment he said but I will let you know if I do. And that was that, or so NHM thought. A few days later Megastrike appeared at NHM's office. What brings you here NHM asked? Megastrike sat himself down and asked what kind of project NHM had been talking about the day they met. When he heard the details he said he would volunteer his service without any salary. Do I understand you correctly NHM asked? You want to work as a volunteer in the lab, is that correct? "Yes", he answered. Aren't you employed at the University of Arizona? No, I'm not he replied. But he gave no further information. OK, send me a *curriculum vitae* NHM told him, and I will let you know in a few days after I get it. Megastrike left. His CV arrived the next day. From it NHM learned he was a disabled veteran from the Vietnam war. NHM too is a veteran from that era and thought, why not give this man a chance? What harm can it do to let him work as a volunteer in the lab? I would be careful Creator advised. Remember this is a learning curve for us. Who knows what might go wrong? Noone could have guessed what did go wrong.

Although he is an exceptionally intelligent individual, a prolific writer and mathematical modeler, as well as a ceasless talker, Megastrike turned out to be a total con. In theory he had heard of everything, but in detail what he thought he knew was totally garbled in his mind. His interpretations of things were ludicrous. He found fault with everything. Eventually he began to insist that he do his own experiments. The things he wanted to do had little or nothing to do with NHM's research objectives. Finally NHM allowed him to go ahead with one of his experiments but insisted that the electrical apparatus he planned to use be enclosed in a box to avoid setting fire to anything on the bench-top should it short out. NHM required that he do his experiment when others were not working at the same lab bench in case something went wrong. Without telling anyone he chose to do his experiment over the weekend. The following Monday NHM found his box in the trashcan. The interior was blackened. The circuit had burned. NHM told him he would have to stop that line of work and focus on something else. NHM kept shifting his responsibilities from one project to another trying to find something he could do well enough to be useful. Nothing worked. Although he talked a good line when you got right down to it he couldn't do a thing. He had no microbiology skills, or even basic understanding of how to work with bacteria. He could not fabricate force transducers using a simple device NHM built because his hand/eye coordination was inadequate. When he tried to do some things on the computer it resulted in what he called, "fatal blue screen", meaning the program had crashed and he lost all his work. One failure followed another. He then took to writing criticisms of things being done by others in the lab. He would disappear for days at a time then present NHM with pages of irrelevant calculations. His distractions soon became too time consuming. Finally Megastrike presented NHM with a piece of statistical information that he derived from one of NHM's data sets using a computer program available in one of NHM's computers. NHM first thought it might be useful but it turned out to be inappropriate. Things came to a conclusion the next day. NHM was in his office going over some work with a graduate student from the Applied Math program who was doing both experimental and theoretical work in his lab. In burst Megastrike. He had in his hand one of his white envelopes that usually took the form of a letter to NHM with his latest advice. NHM had had enough by then so he handed it back to him and asked him what he believed the last thing he had presented was supposed to mean. Taken by surprise, Megastrike could not answer. He became flustered, and stomped away. NHM found a letter from him in his mailbox the next morning. In it he informed NHM that, "If you wish me to continue to work for you, you can buy me a sixpack of Bass Ale to calm my jangled nerves. Otherwise, it't (sic) only a hobby. Fuck it! ". And he signed his name below. Guardian gave the instructions. You have gone out of your way for this individual but the time has come to get him out of your laboratory as quickly as possible. He is a dangerous person and a risk to all concerned. How right Guardian was.

NHM drafted a letter and took it to the department's administrative assistant to make sure that he had the right to terminate a volunteer. Yes nowadays such things must be written correctly from the legal perspective, not something a principal investigator can assume responsibility for in the context of his own research program. Once NHM was assured it was within his rights to do so, he forwarded the vetted letter to Megastrike. In it he thanked him for the work that he had done and for his efforts. It then asked him to move on to other pursuits instructing that he remove all his things from the laboratory and return all keys. Finally it wished him success in his future pursuits. In spite of the careful wording of the letter it did not solve the difficult problem it was meant to. On the contrary, it exacerbated the conflict and led to a dangerous situation. Megastrike could not accept being dismissed. He was determined to retain his status and so he dreamed up a unique way to do so. His method was to write two letters, the first to the Office of Research Integrity, an office established within the Department of Health and Human Services to deal with fraud in science, the other to the "Vice President in charge of Research" at the University of Arizona. Both letters are reproduced below.

United States Department of Health And Human Services
Directors, Office of Research Integrety
5515 Security Lane, Suite 700
Rockville, MD 20852

Dear Sirs,

I am a Service Connected Disabled Veteran of the Vietnam Era who works for Dr. Neil H. Mendelson at the University of Arizona Department of Cell and Molecular Biology in the Life Sciences South Building on the main campus of that University.

Dr. Mendelson asked me to work for him in evaluating the motion of bacterial colonies in cultures that he had established in his laboratory. As I worked with the programs that he had bought to do the analysis, I found them to be defective. I reported this to Dr. Mendelson and wrote a small paper on the subject (marked in the enclosed mss.) His reaction was to cover up the discrepencies between promise and performance and to instruct me to tell no one outside his lab.

The problem resoved down to this: Though velocity is a vector quantity having both magnitude and direction, the program threw away the direction value and reported only the absolute value of magnitude. Thus, it is impossible to calculate acceleration, the rate of change of velocity, another vector quantity, from the "tracker" program used by Dr. Mendelson. Dr. Mendelson's answer to this was to hire a chinese Engineer, who hardly speaks English, to calculate acceleration. I looked at her results and sure enough: all values were positive and there were no indications of direction.

If Neil Mendelson is dishonest about data reduction, he may be dishonest in other matters. B. Subtilis is suposed to grow well on Hay Infusions, according to Burrow's and the Bergy's Manual of Determinitive Bacteriology, 7th Ed. We are growing it on Triptose-Beef Extract with added NaCl. That is: we are growing it on the wrong medium. Now, there are several types of Motility that have been identified to me with his strains: 1) gliding, like Myxobacter sp. 2) Swarming, like Proteus sp., 3) coiling, like Antraxis sp. And finally swimming. This bacterium is doing too many things to be the same bacterium in all cases. If there were Gram staining materials in the lab, I could determine whether we were dealing with Proteus or not, as

it is Gram Negative; whereas, B. subtilis is Gram Positive. There are no staining materials in the lab at present. Simply put: this organism is evolving too fast to be one organism.

I also doubt his judgement on Biological Warfare. He maintained that it took a building full of equipment to micronize Anthrax sp. spores. Apparently, what happened in Washington, Florida and New York was that someone got hold of a glove box, a respirator, a lot of penicillin a morter and pestle and some sand and ground up a dried, potato agar sporulated sample of Antrax bacillus and stuffed it into envelopes and mailed it. I suspect a domestic origin for the stuff, done probably by Iranian Terrorists (There is an old Persian proverb: "The mills of God turn slow, but they grind exceeding fine!" It probably took some fanatic a little while to do it, but do it he did!

Finally, I came in contact with some of the culture medium that had bacterial colonies growing in it, due to improper disposal methods in the lab. I developed a lesion on my hand that I cured with neosporin, as a result. If we were dealing with Hay infusions and B. subtilis, this shouldn't have happened. I was unable to wash my hands in this incident, because there was no soap in the lab. I have since supplied the lab with Dial anti-bacterial soap in a dispenser to avoid this problem in the future.

The software programs and the materials were paid for with an NIH grant. There is a lack of proper oversite and direction in this laboratory and in this department of Cell and Molecular Biology. Dr. Mendelson is a fine experimentor, but he does not know how to reduce his data and stoops to subornation and deception to hide it: velocity and acceleration are vector quantities and have both magnitude and direction. No one who has ever calculated an asteroid's position in orbit to a gnat's eyelash can ever forget it. /

This is the letter he sent to the University of Arizona:

To whom it may concern:
At the Office of Academic Integrety
Co/Vice President in charge of Research
Room 601, Administration Building
University of Arizona
Tucson, ARIZ 85716

Dear Sir,

I am a whistle blower, blowing the whistle on Dr. Neil H. Mendelson of the Department of Cell and Molecular Biology for producing fradulent data using a computer program which he calls "Tracker." The "Tracker" (Image Analysis) program is at fault and I have advised him of this. It produces 19 columns of data for 19 time frames of image data. It should produce at least 38.

There are two ways of reporting velocity data in two dimensions, both involve ordered pairs of data (V_x, V_y) or (V_r, V_{theta}). (See: <u>Astromechanics</u> by Peter Van De Camp) The "Tracker" program only reports $|V_r|$. I advised Dr. Mendelson of this problem, but he chose to cover it up, hiring a Chinese PhD in Engineering to calculate acceleration from this data. It is unfortunate, but you can not calculate acceleration from $|V_r|$ alone. Dr. Mendelson moved me out of my desk to set up a high speed camera that he had bought and which should have been the property of an Academic Consortium. He instructed me not to tell anyone, just as he had instructed me not to tell anyone about the program, "Tracker." Lili, the Chinese Engineer showed me her work. All values of acceleration that she calculated were positive. Nothing decelerated in her images! My guess is that she did not understand what she had produced. Dr. Mendelson seemed happy with her work, while he glowered at me when I brought in a power supply for a 486i computer that he was going to allow me to use to run a F-test program that I had written to run on some data I had calculated for him, which he claimed showed "no difference" in statistics between a random sample and a decidedly non-random one. He never did give me the data to work with, though he had it for over a semester. Nor would he believe me that the random one seemed to follow the <u>cumulative</u> normal distribution, a cubic or quintic shaped curve in the books. The "F" test would have shown diffenences in variance in the several plots.

Dr. Mendelson's pattern of deception and disenfrancisement seems to be a conscious manefestation of the way he conducts research. His lab is a mess. It lacked soap to wash one's hands with until I went to Walgreens and bought some. Disposal of cultures led to an infection, in my case, which I cured with neo-sporin.

Worse, there was no cresol (lysol, pinesol), phenol or Sodium Hypochlorite to clean the benches. I know my bench was often contaminated with spore forming bacteria which the ethanol provided would not take care of. Ethanol is a drying agent with little killing potential for infective agents that form spores. Liquid culture medium was placed in petri-dishes without a gelling agent and innoculated with spore forming bacteria. More than once, these cultures spilled on the bench, contaminating the work area and the workers. It seemed almost inevitable that infections would occur. I respectfully suggest that Dr. Charles Gerba of Soils and Microbiology swab the benches and see what grows!

I have been shown 4 different motility modes for Bacilus subtilis in his lab, modes suggestive of Myxobacter, Proteus, B. anthraxis and B. subtilis, with the claim that all organisms were B. subtilis, yet there was no way to check for contamination other than smell! I could find no Gram Staining materials in the lab to check for Proteus. Shortly before the summer, Dr. Mendelson reported to me a stomach infection which he supposed to be Helicobacter pilori. I wrote him in Oregon that it was more likely a mixed infection of non-pathogenic H. pilori and Proteus morgani, which would have the same effect. From that point on, he began deriding me as "The Big Microbiologist." However, his "ulcer" did appear to clear up. He disposes of bacteria grown in his chambers for tensile force measurements by dumping them down the sink.

I believe that these actions constitute a violation of the Ethics in the Workplace policies of the University of Arizona. I believe further that he is teaching his students bad work habits with bacteria and is in the process of falsifying data to do with ac

cleration. Academic Labs are notorious for their sloppyness, However, his tops anything I have seen for the carelessness with which bacteria are handled.

I would like "Whistle Blower" protection and to be reinstated in the lab. Dr. Mendelson has himself to blame for this complaint. Had he dealt with me fairly and honestly it never would have happened. I am only a High School Teacher, but I do remember Dr. Sinclair carrying his secpter (an innoculating loup) through the lab to sterilize it after use, even though it was contaminated with a non-pathogen. Dr. Mendelson is working with pathogens. It is necessary to be careful with them.

Neither of these letters came to NHM's attention until early in January when a University of Arizona police officer, a Mr. G. Ewer, came to speak to him in his office. Ewer informed NHM that a letter had come to the University from the Office of Research Integrity and asked if he thought there might be a risk of violence by "Megastrike". NHM told him he could see no potential threat but he had not yet gotten his copy of the correspondences that were the cause of Ewer's concern. Later that day when NHM's copy of the entire file arrived, he did see that there was a serious problem. Nothing in the initial letters was of concern, but by the time the university informed NHM in January 2003, the Department of Health and Human Services (HHS) had gotten additional letters from Megastrike which clearly demonstrated that a danger existed. HHS sent NHM a copy of the material that frightened them at the same time they did to the University of Arizona administration. This provided NHM with the information he needed to act in a timely manner. One layer of bureaucracy had been eliminated. Here is the HHS letter retyped for clarity of reproduction:

DEPARTMENT OF HEALTH & HUMAN SERVICES

Public Health Service

Office of Public Health and Science
Office of Research Integrity
5515 Security Lane, Suite 700
Rockville MD 20852

CONFIDENTIAL/SENSITIVE

December 31, 2002

Dr. Richard Powell **FAX (520) 621-7907**
Vice President for Research and Graduate Studies
University of Arizona, 601 Administration Building, P. O. Box 210066
Tucson, Arizona 85721

Dr. Thomas Hixon
Associate Vice President for Research

University of Arizona, 601 Administration Building, P. O. Box 210066
Tucson, Arizona 85721

Dr. Neil Mendelson, Professor
Department of Molecular and Cellular Biology
University of Arizona, Life Sciences South Building, P. O. Box
210106
Tucson, Arizona 85721

RE: D10 2405

Dear Drs. Powell, Hixon, and Mendelson:

The Division of Investigative Oversight (DIO), Office of Research Integrity ((ORI) received the attached letter from XXXXXXXX in Tucson, Arizona of whom you are already aware.

While there are a variety of ways that one might read his letter regarding "bottle of cyanide," "disposing of the whole campus of the U of A and a bit of the surrounding town," and "murder," this letter is of sufficient concern to ORI staff, regarding the health and welfare of him and your employees, that we wanted to ensure that you had seen it and that you had the opportunity to inform the appropriate authorities.

Sincerely,

Alan R. Price, Ph.D.
Director, Division of Investigative Oversight
Office of Research Integrity

Here are the letters that Megastrike sent to HHS that raised their concerns:

December 24, 2002
7:58 AM

Department of Veteranary Sciences
University of Arizona
Tucson, ARIZ 85721

Dear

I have a mystery that I would like to see cleared up in the New Year: "What was Dr. Mendelson doing with a half a gallon jar of Sodium Cyanide? and what happened to the other half gallon?" This compound is used in turning off respiration in facultative anerobes (that is shutting down the krebs cycle through inhibition of the cytochrome oxidase chain) so that one can study mechanisms like glycolysis (a very inefficient way to derive energy from glucose) with out interference of oxygen carrying heme compounds like Cytocrome C and Haemoglobin. As human beings use these compounds to derive energy from their food, Sodium Cyanide, NaCN, is a very effective poison against humans. Dr. Mendelson works only with Bacillus subtilis, which is wholely aerobic and had no use for the Sodium Cyanide, at least no legitimate use.

The quantity is amazing! Sodium Cyanide is used in milliequivalent amounts to turn off respiration. Surely a pint would last any legitimate University investigator a life-time. When Cyanide compounds react with Haemoglobin, they produce a blue pigment (Cyano-haemoglobin) due to the reaction of the Cyanide radical with ferous iron, found in the protein. The blue pigment turns the lips and flesh under the fingernails a dull blue (prussian blue) and the condition is known as "cyanosis" (a simmilar but less pigmented form occurs when people stop breathing.)

I have already supplied to you a little paper in how to decontaminate Sodium Cyanide poisoned drinking water. I suggest that an excess of iron II sulphate be added to the water with the addition of calcium hydroxide (to scavange the sulphate radical) and the water be cleared by floculation.

Now, I may have run into this bottle of Cyanide

before: Norval A. Sinclair had a bottle like it in his lab in 1975 or so. I complained to it to my Psychatrist, Dr Hubert E. Wuestoff; Hub had been head of the Hospital at Davis Monthan Air Force Base in Tucson before retiring to private practice. In any case, I was feeling suicidal because of mal treatment by a girl friend and did not appreciate having a bottle of cyanide readily available in an open lab at my disposal. The bottle was removed from the lab at that time and apparently given to Dr. Mendelson, who has a locked lab to keep it in; however, my experience is that in Dr. Mendelson's lab there is at least one practical joker and the idea of having Sodium Cyanide at such a person's disposal makes me uneasy. There is appoximately enough Cyanide in that bottle to dispose of the whole Campus of the U of A and a bit of the surrounding town. It should be disposed of as I indicated in my earlier letter, not be left sitting around waiting for a practical joker or terrorist to make use of.

Cyanide is only one of the many metabolic poisons on campus. Dr. Klaus Brendell of Pharmacology gave me Sodium Azide and PMSF to play with in the lab without any warning on their danger, while I was a volunteer Research Assistant in his lab. As Sodium Azide is about 30 times as potent as Sodium Cyanide, this was hazardous behavior on his part and probably constitutes negligence.

Finally, the practice of mouth pipeting reagents in a biology lab was not mentioned in the Ethics Committee's report, but Dr. Mendelson required us to mouth pipet solutions into petri dishes in order to make up culture media. This is in violation of NIH guidelines for any Biology lab. One piece of instruction rings out in my mind from my Biochemistry classes prior to the issuance of the NIH Guidelines: "Don't mouth pipet the Cyanide!"

Merry Christmas

[redacted]

This letter suggests that the claim to be a "Service Connected Disabled Veteran" may in fact be based upon a psychiatric disability. It certainly reveals suicidal tendencies that should have prevented him from working in any laboratory where he would have access to dangerous chemicals or instruments. Unfortunately the University of Arizona Personnel office has no mechanism in place to identify such individuals. Had NHM known that this individual was potentially suicidal beforehand he would have never been permitted into his laboratory in any capacity. A copy of the second letter he sent to this same person earlier that same day is reproduced below. In this letter the focus shifts directly to NHM as the source of potential death.

December 24, 2002
2:33 AM

Department of Veteranary Sciences
University of Arizona
Tucson, ARIZ 85721

Dear ,

I told you that Dr. Mendelson kept his culture medium tubes for 2 to 3 months in steel capped tubes so that they evaporated mightily before they got used (They might evaporate to 2/3 their original volume in the process of waiting.) This compromised the defined nature of the medium; however, there was an offsetting error: He would add a full litre of Milequ water to his measured solutes rather than QSing it up to a litre, so that one would wind up with more than a liter of solution to autoclave.

Thus his claim to use a defined weight per volume solution to grow Bacillus subtilis is all wet! We are not talking about plus or minus 5%, we are talking about plus or minus 33%. To my mind what he is doing is not Science but a form of witchcraft: "Bubble! Bubble! Toil and Trouble!/ Fire burn and caldron bubble! When shall we three meet again?/ In fire, lightning or in rain?/ When the hurly burley's done!/ When the Game is lost and won!/ But here comes Dr. Mendelson! (I mean: "Here comes McBeth!")"

Have a merry Christmas, remembering that you do not treat lab workers to a lot of invalid results as the result of your sloppy chemistry ala Mendelson. If there was one thing I learned as a Research Assistant in Pharmacology: You QS upto volume to make solutions, you don't just pour in a measured amount. Dr. Ray Duhammel may have been a little on the dishonest side in his publications, but at least he knew and taught how to make solutions correctly.

Best to you,
Merry Christmas

Finally there is a poem that he wrote entitled, "Attempted Murder?" shown below.

Attempted Murder?

When I was cleaning out the lab
After doing experiments with paramecium
I asked Dr. Norval A. Sinclair
What I should do with a bottle
That said it contained Ethanol
"Drink it!" he said.
That went against standard lab practice:
One doesn't eat or drink in the lab,
So I asked again what I should do with it
"Drink it! It's Alcohol!" was the reply.
I told him I wasn't going to drink it
What should I do with it?
"Pour it down the sink," he said in desperation.
It wasn't until years later
I remembered what was in the bottle:
It was "Gluteraldehyde." And enough of it
To kill a platoon of men deader than a doornail.
I had gotten it a couple of months earlier
From Dr. Timmothy Ferris, who used it
As a fixitive in electron microscopy.
There was no proper bottle for it so
Dr. Sinclair had had me put it in an Ethanol Bottle.
Had I drunk it, I would not be writing this today,
And another unfortunate lab accident
Would have claimed another fool;
It is not wise to eat or drink in the lab:
Your life may depend on it completely!
The wildlife in the holding tank died instantly
On contact with that stuff:
"First they had Paramecium,
Then they were dead," they said!
Was it intentional on Dr. Sinclair's part?
I don't know, but had I drank it as he said
They would have sectioned my gut with a microtome
And inspected the sections with a microscope.
Please don't eat or drink in the lab,
It makes one hell of a mess for the corroner to clean up!

Megastrike must have known these writings would elicit an immediate response on the part of the Office of Research Integrity at the Department of Health and Human Services. They did. The University of Arizona took no serious action of any sort to protect anyone from him however. This was quite surprising to NHM in view of the fact that the University had just suffered a terrible incident at the College of Nursing in which a disgruntled student murdered a number of professors and administrators. Instead our Vice-President for Research focused on having an investigation made to determine whether NHM was indeed guilty of the accusations made by Megastrike. A committee was formed to investigate the charges. In preparation for appearing before them NHM drafted the following response and distributed it to all relevant parties.

The University of Arizona
Department of Molecular & Cellular Biology
(520) 621-7560
FAX (520) 621-3709

Life Sciences South Building
1007 E. Lowell Street
P.O. Box 210106
Tucson, Arizona 85721-0106

January 17, 2002

MEMORANDUM

TO: All parties dealing with allegations made by [redacted]

FROM: N. H. Mendelson, PhD, Professor of MCB

RE: Response to claims and correction of facts

Concerning the letter sent to the NIH office of Research Integrity

Response to paragraph 1. [redacted] is not now and never was employed by me. He worked as a volunteer in my laboratory from October 30, 2000 to November 7, 2001. He presented himself as a volunteer several days after I asked him at a chance meeting one day on campus if he knew of any students who would be interested in doing a research project in my lab. During the time he was affiliated with my research laboratory his behavior became increasing erratic and oppositional. I eventually discovered that he has a history of writing letters to people in positions of authority in which he attempts to discredit individuals of accomplishment. A perfect example is an unpublished "book" that he has written focused upon Albert Einstein's error. In it there were copies of numerous letters he sent to people arguing why Einstein was wrong and he is right. The people to whom he sent them were the Queen of England, high ranking US government officials, and so forth. In this "book" he cast himself as a lowly high school teacher who knows more about the issues than Einstein did. In a similar vein I have copies of letters he sent to Senator John McCain informing him of why the Excalabar weapon proposed by Edward Teller will not work, to the director of the National Science Foundation telling her why NSF's initiative in biomathematics will not work, letters describing why there is something wrong with the work of students in my lab, Darshan Roy and Catherine Ott in particular, why our commercial statistics software programs are no good, why we need to determine if there are rays coming from the fluorescent lights in our lab that cause the results we obtain, and on and on. I have a copy of his writings on anthrax assuring people at the Centers for Disease Control and Prevention that "the press has hyped the dangers from Anthrax out

mcb

of all proportion". In view of all this I am not surprised therefore that he has attempted to discredit my work as well particularly in view of the fact that I terminated his affiliation with my lab.

Response to paragraphs 2 and 3. One of several projects pursued in my laboratory is the study of motions in multicellular bacterial forms. often criticized the approach taken in our work and refused to do what he was asked to do. For example, he found fault with a tracker program we used to study velocity and direction of multicellular forms because he believed we should be measuring acceleration instead. That was not our objective; nor is it 's prerogative to direct our research. The tracker program was perfectly suited for what we needed to measure. In addition, the claim that Dr. Chen (an American citizen of Chinese descent) was hired to calculate acceleration is totally incorrect. She has never used the tracker program in her work. Her first assignment was to analyze data concerning velocities and to calculate forces from the measurements. s claim of "cover up" and instructions to "tell no one outside" about our work is pure fabrication.

Response to paragraph 4. There is a long publication record that attests to my honesty in science, not dishonesty. 's comments about growth medium composition, e.g. hay influsion, are totally uninformed. There are many different kinds of media in which *Bacillus subtilis* can be grown. is not qualified to tell us which medium to use for our work. In fact the growth medium that we use, I adopted over twenty five years ago and have successfully used to culture *Bacillus subtilis* precisely because it facilitates growth in the multicellular form that we study. Indeed I have discovered many unusual things that *Bacillus subtilis* does and am well know for having done so. Many other scientists have independently confirmed the correctness of my discoveries. Mr. s assertion that more than one bacterial species is growing in our cultures is incorrect speculation. Having worked with *Bacillus subtilis* for almost forty years, I know what is and what is not *Bacillus subtilis*. Discovery of the great diverse potential of this single bacterial species is no indication that "this organism is evolving too fast to be one organism". This is pure nonsense.

Response to paragraph 5. Biological warfare has nothing to do with what we do in my laboratory. Paragraph 5 suggests that Mr. is motivated by other factors to submit his criticisms.

Response to paragraph 6. In my laboratory all required safety precautions are taken with regard to disposal of live bacterial cultures even though the organism we study, *Bacillus subtilis* , is a well know non-pathogenic model organism, found widely distributed in soils, and eaten in large volume in both Japan and France. [redacted] claims a lesion on his hand developed as a result of contact with a bacterial culture in my laboratory. He provides no medical diagnosis to support his assumption that a laboratory contact was responsible for his skin lesion if indeed he had one. There has never been a laboratory acquired infection in the nearly forty years that my laboratory has existed. Finally, antiseptic bar soap has always been available in my laboratory.

Response to paragraph 7. Mr. [redacted]'s contention that neither I, as the PI, nor the department setting of my laboratory provide proper oversight of the funds awarded by NIH to me, is something that he is not in a position to judge. I am appalled that an individual who volunteered to work in my group and to whom I extended many privileges and opportunities should make the outrageous claims that Mr. [redacted] has made. Mr. [redacted] appears to be very angry because I asked him to leave my laboratory. I did so, when it became clear that his presence was a total distraction to the ongoing work of the laboratory and his efforts were consistently unproductive or peripheral in their direction. The final straw came when he wrote a belligerent letter to me making it clear to me that he was unsuited to be a member of my group in any capacity.

Concerning the letter sent to the University of Arizona

Response to paragraph 1. No fraudulent data have ever been produced in my laboratory. The "tracker" program used in our laboratory was developed in the Medical School at the University of Rochester to measure the direction and velocity of objects such as motile cells. We have applied it in the study of motions found in multicellular bacterial forms.

Response to paragraph 2. a. Mr. [redacted]'s focus on the improper use of the "tracker" program to measure acceleration is misdirected. He does not appear to understand that the program was never intended to be used to measure acceleration, and we have never used it for that purpose. The relevant information for our work is merely the direction and velocity of multicellular bacterial movement. Tracker is well suited to this purpose. Furthermore the measurement of velocities we have made with the tracker program have only been used as guidelines for our further analysis. None of the velocity data have yet been published. Only data dealing with direction of movement obtained with tracker are now in press.

b. The hiring of Dr. ████ (an American citizen of Chinese descent) was not intended to "cover up" anything. She is an accomplished engineer who brings talents to our projects in areas dealing with the measurement of forces created by bacterial macrofiber supercoiling. Her only work dealing with motions of multicellular bacterial forms of the type Mr. ████ dealt with concerns estimating forces. None of her work has anything to do with the tracker program. I am aware that Dr. ████ did at one time calculate acceleration rates without using the tracker program. Those data were never used because I determined that they were not relevant to our objectives. Mr. ████ wanted to determine acceleration using the tracker program but was unsuccessful. He could not in fact use the program to measure anything. I believe the issue of measuring acceleration assumes an importantance to Mr. ████ because he could not obtain the measurement that Dr. ████ did. The rate of acceleration has never been an important aspect of our project however.

c. I did use Mr. ████'s desk as a platform for new equipment in view of its proximity to other instrumentation. By the time I did so however Mr. ████ had virtually ceased doing any work the laboratory. Instead he did his theoretical work at home. Shortly thereafter I asked him to leave the laboratory altogether. The high speed camera to which he refers was purchased specifically to support a main objective of my NIH research grant. Its use is dedicated exclusively to the needs of our project. We are not presently in a position to provide time on this machine for investigators outside of our group. This is what Mr. ████ refers to in his comments about an "Academic Consortium" and instructions not to tell others that we had a high speed camera in operation. His reference to not telling anyone about the tracker program is totally false.

d. ████ did indeed bring a computer program to the lab that he claimed could perform statistical analyses. There were two problems with it however: i. we already have a program installed that can do the calculations, and ii. when Mr. ████ tried to run his program he caused our main computer to crash. He came to me and told me we had "fatal blue screen" on the main computer, by which he meant he could no longer get the computer to operate. Once I restored the computers function I vetoed his request to run his program on an older model 486 computer that we keep running because we have some data sets still active stored in that computer that are difficult to transfer to any of the newer computers. It is important therefore that we keep the 486 functional until all the data sets it houses have been analyzed and published. In addition I forbade him for running his program in any of our computers.

e. The data sets to which Mr. ████ refers have been properly analyzed in consultation with both a probabilist and an engineer/mathematician with whom I have collaborated for twenty years. The publication containing this analysis will appear soon in Pub Med Central for those who want to examine it. Mr. ████s F-test is not the appropriate way to show significance in this instance. Once again his approach has been counterproductive.

Response to paragraph 3. What Mr. ████ sees as disenfranchisement is simply a principal investigator's means to keep control of the research done under his direction and to ensure that appropriate expertise is applied. Deception is no part of the agenda.

Response to paragraphs 4 and 5.

a. The claim that my laboratory "is a mess" is a judgment made by someone unfamiliar with laboratories of this type pursuing the most original research all the time.

b. Not only was there soap in the laboratory before Mr. ████ bought his bottle of Dial soap, but in the cabinet immediately below the sink there are and have always been both bleach, and benchtop disinfectants that he apparently failed to notice or use. So far as disposal of cultures is concerned, there are no large cultures produced as there would be in

5

conventional bacteriology. Everything is done on a microscale in individual Petri dish cultures that are discarded when no longer needed in official biohazard disposal bags. Mr. [REDACTED]'s claim that he obtained an infection from a disposal culture is not supported by any medical diagnosis. It is merely an assumption. I was never informed by him during his tenure in the laboratory that such an infection existed. Mr. [REDACTED]'s claim that ethanol is an inadequate means to kill the cells that we work with in wrong. We do not work with spores. Only vegetative cells that are rapidly killed by ethanol are cultured. Furthermore his contention that *Bacillus subtilis* is an infectious agent is also wrong. It is to be noted again that this is a common soil organism eaten in tofu in Japan and taken as a digestive aid in France.

Response to paragraph 6. a. Mr. [REDACTED]'s claim that there is no way to check for contamination in our cultures other than by smell is nonsense. We grow *Bacillus subtilis* in an unusual macrofiber form that results in clear cultures containing bacteria only in the form of white threads. Any bacterial contamination in the growth medium would rapidly produce a cloudy culture, immediately apparent and easily killed and disposed of.

b. Mr. [REDACTED]'s comments about my health condition are truly unwarranted. He is neither physician nor diagnostic microbiologist. His assumptions in this regard are typical of his willingness to assume a position of knowledge beyond the scope of his education and expertise.

Response to paragraph 7. I have worked with *Bacillus subtilis* not only for thirty three years at the University of Arizona, but also, as a postdoctoral fellow in the Medical Research Council's Microbial Genetics Research Unit, Hammersmith Hospital in London, as a Visiting Professor at Institute Pasteur in Paris, as a Visiting Scientist at the University of Lausanne in Switzerland, at Cambridge University in England where I established a laboratory in the engineering department as a subcontract from my NIH grant, at the University of Rochester Medical School in NY, at the University of Maryland in Catonsville, and during the Viet Nam war as an officer assigned to the Army Biological Laboratories at Fort Detrick in Frederick Maryland. In addition I am a fellow in the American Academy of Microbiology. Mr. [REDACTED], who claims to be a high school teacher, although his cv shows very little evidence thereof (between 1968 and the present he lists student teaching in 1973,

1982-83 teaching in Phoenix Indian High School, and 2 months teaching in Mexico in 1988) is in no position to tell me or you about how to handle *Bacillus subtilis* in a responsible manner.

Response to paragraph 8. Mr. [redacted] wishes to be reinstated in my laboratory. That is totally unacceptable to me. His claim that his complaint would never have been lodged had he been dealt with "fairly and honestly" suggests a motive of revenge for his dismissal from the laboratory. If he had a genuine concern about safety it would be hoped that he would have reported it long before now. The idea that I am working with pathogens reflects an inability of Mr. [redacted] to know what is going on about him. *Bacillus subtilis* has been used as model organism precisely because it is definitively non-pathogenic.

In sum, there is no validity to any of the claims made by Mr. [redacted] in either of his letters of complaint. This is a nuisance complaint and should be dealt with as such.

Distribution:
Dr. Alan Price, NIH, ORI
Dr. R.C. Powell, VP Research UA
Dr. T. Hixon, RIO, UA
Dr. J. Ruiz, COS, UA
Dr. D. Brower, MCB, UA
Dr. S. Wilson-Sanders, Ethics and Commitment Committee, UA
Mr. M. Grushka, Biosafety Officer, UA
Mr. H. Wagner, Risk Management, UA

This memo was meant to set the record straight. When NHM appeared before the committee he brought many ancillary documents pertaining both to Megastrike and to himself including details pertaining to his own military service to the rank of Captain in the US Army and his recognized accomplishments as a microbiologist. After the investigative committee submitted its recommendations the Vice-President for Research "dismissed all charges against Dr. Mendelson" and said that, "I consider the matter to be closed". He sent NHM a letter on March eighth 2002 in which he stated, "I am gratified to see that the Committee's findings were negative and regret any inconvenience and grief this process may have caused you". It was unfortunate that Dr. Powell had no idea of who NHM was in the first place. The Vice-

President treated this matter from the very start as if NHM and Megastrike were on a par in terms of their credibility. It was a perfect example of how little the central administration knows about the faculty, or the history of the institution. There was a time long ago when the founding Vice-President for Research, A. R. Kassander, asked if NHM would provide him with a figure showing the detailed structure of a macrofiber, and when he received it he had it framed and hung in his office. On a recent visit to the same office NHM noticed that his old photo mosaic was gone. No one working there presently knew where it might be or even what it was. So much for longevity.

No action was ever taken by the University to limit Megastrike's campus access, to warn other faculty or to directly protect NHM or others in his laboratory. Here is the last correspondence NHM has seen from Megastrike:

OFFICE OF

Directors
Office of Research Integrety
NIH
Bethesda Md

Sirs,

An honest difference of opinion between me and Dr. Niel Mendelson at the University of Arizona?

Notice the "I think not!" above the blacked-out signature block above.

What a misery our volunteer turned out to be, Creator said. I feel responsible for it having raised the suggestion in the first place. I only wish now that I had kept my mouth shut. It will take me a long time to live this one down. Two years later however Creator was fully back on his feet. He planned an outlandish fairwell talk NHM gave to the Molecular and Cellular Biology Department. What it involved will have to be held for another time, perhaps another story of life after retirement or some such tale. The important thing here is just that even after a major mistake, Creator bounced back and thus the chaos that NHM fostered everywhere lived on beyond Megastrike.

10. Three strikes and you're out

NHM was very happy to finish his perfect mistakes career at Arizona even though there was a lot of humor in it and it served his essential needs very well. At Arizona he managed to do some praiseworthy research. In the decade of the 1990's quite a few scientific commentaries were written about his discoveries and inventions. He included some of them in a book titled, "Collected Papers of Neil H. Mendelson" an archive of all his published works. Publications are however only the tip of the iceberg of what NHM had found, which is probably the case for all productive scientists who write their own papers. What happens to all the other findings when one closes down shop? In most cases they are probably lost. NHM was fortunate to have been invited to contribute his data books and other writings to a repository where anyone can have access to them. They are now housed in a new bomb shelter-like building in Laramie, Wyoming known as the American Heritage Center. To find his things look for, Mendelson's papers, number 9361. Yes NHM was very fortunate in many regards but this escaped detection at the University of Arizona. Not only did his work go against the grain there but his culture did too. Those scientists who knew him at the early stages of his career and who could have promoted him never understood why he chose to go to Arizona in the first place. It was not the kind of institution they expected him to end up working at, and, to be frank, could not see the value of promoting him in that setting. Get yourself to Harvard as fast as you can, one of them advised him, but nothing like that was to be. For the same reason that he turned down an offer from the University of Texas at Austin, he really couldn't think about leaving Arizona. Just do your work and put it out of your mind is what Creator told him to do and that became his goal. Now that he has done so he is not certain that he could have done better anywhere else. The vast majority of what is in his Collected Papers book really doesn't match the culture at any institution that he is aware of and he himself probably wouldn't have fit better elsewhere even if it did. NHM has thought long and hard now about why that should be the case. Although he may not yet fully understand the predicament of his life, NHM believes that if you trace the history with him of what led to all his perfect mistakes you will come to the conclusion as he did, that it all makes perfect sense. Nothing else could have been the outcome.

To tidy things up Observer suggests that we first finish the story about working at Arizona. Three strikes in the lab helped determine precisely when research came to an end. But that still left a lot of things to do: retirement, disposing of all the research tools collected over 39 years, getting everything sent to the archive, and making sure to keep everything needed for the writing of the three books. And there was one remaining obligation that NHM could not forget, he was assigned to organize one of the weekly MCB department Teas. This obligation fell upon each faculty member usually once each year. What had to be done was to provide food and snacks for the department community to devour in advance of attending the department seminar, usually a talk given by an invited scientist from another institution. Some of these semi-catered affairs were rather elaborate, others much less so consisting of hand crafted cookies or the like. The Tea routine was something akin to a British tradition NHM was familiar with from his days in London and Cambridge. On the day NHM was assigned there was no department seminar. Without having to hurry off to a seminar the Tea on those rare speakerless days generally ran longer than usual ending only when everything had been consumed. Why not take advantage of this extra time Creator suggested? How about a Tea they will never forget Performer urged? Yes, but what should it be NHM wondered? Well it took some thought to develop the plan but this is what it ended up being. A three-fold event was planned, kind of like a three-ring circus. In one ring there would be a small jazz band playing background music throughout the event. In the second there would be an elaborate desert table with cakes, ice cream, and such things. In the third ring would be

NHM with a cotton candy making machine producing the spun product for everyone, and instructing those who wished to make their own how to operate it. NHM's colleague Jim Karrer, principal bass in the Tucson Symphony put together a trio for the performance. NHM's wife Joan organized ring two, and also found a carnival supply company from which NHM could rent the machine and purchase the colored sugars used to make the product. Everything worked! The place went wild. Word spread quickly to other departments and they flooded into the building. People appeared with cameras to capture the response and the true circus-like atmosphere that enveloped the place. Here are some examples:

NHM had alerted his colleagues in math and physics and some of them managed to get over to see it in action. Mike Tabor, his collaborator from Applied Math was tutoring a graduate student when NHM called and told him he must not miss it. Tabor decided to bring the graduate student along with him. When the elevator opened on the fourth floor they stepped into mayhem. Michael later told NHM that he turned to the shocked student and said in his very British way, "Oh yes, this is Molecular Biology on Thursday afternoons". Based upon the number of candy cotton holders that we went through it appears that about 120 people attended. Usually there were 30, perhaps 40 on a good day. Afterwards NHM received several e-mails from people who were there, and others came to tell him how thrilled they were with the whole thing. Danny Brower suggested that NHM had raised the bar significantly. Indeed NHM would have hated to be the one who had to put on the following Tea. Things are back to normal now. No trace of the "Cotton Candy Club" as Ann Webb the woman who took many of the pictures called it, can be found today. It appears to be one of those things that happen once.

On the day that NHM officially retired, July first 2004 and became "Professor Emeritus" Sam Ward was still in the position of Acting Head of MCB. That was fortunate for NHM. Sam appreciated the kind of science that he did, knew of his affiliations with the applied mathematicians and physicists, and Sam helped arrange for NHM to be able to occupy his lab space as well as his equipment after retirement until he finished a few last projects. In addition Sam sent NHM the first of three "official" letters of congratulations upon retirement that he received from the university administration: one from Sam acting as department head, another from the dean of the college, and the last from the president of the university. All are kind and thoughtful letters but when comparing them as you might expect, as you go higher in the chain of organization you notice there is not much really specific that can be said. Neither the dean nor the president really knew anything about NHM's career at Arizona or his contributions in science. The dean, Dr. J. Ruiz, got a small glimpse of what NHM was doing in an unusual way. He appended himself to a group of visiting dignitaries NHM was asked to show through his laboratory a year or so before he retired to illustrate the kind of research that was being done in the College of Science. Perhaps he thought NHM wouldn't recognize him in the crowd. But he did. Afterwards NHM was told that Ruiz became fired up with enthusiasm about the fact that NHM had been able to measure the force of supercoiling directly. Perhaps so, but he never breathed a word of it to him. And in the context of everything that had been achieved over the duration of NHM's tenure it was in truth a tiny part of the whole. Nevertheless the dean's insight was orders of magnitude above that of the president. The president, Dr. P. Likins, knew virtually nothing about NHM or about the history of the institution for that matter. How could he have known that NHM received the first NIH Research Career Development Award given to anyone at the University of Arizona, that he fostered the development of the Biomathematics Program, that Biomathematics received the first IGERT award from the NSF to the University of Arizona? How would the president know that NHM helped organize the Cancer Biology Research Program, and the Biomechanics Program in the Medical School, that he served as a Department Head for four years, that he played in the summer Opera Orchestra at the Music School for several years, that he .... and so on? No he couldn't have known any of that. What he did know was that NHM was a name and a number on a faculty budget line, what NHM looked like, and that's about it. Wouldn't one expect more responsibility on the part of the CEO to at least know something about the senior professors? Perhaps not. Institutions like the University of Arizona are large enterprises and even though in the modern age computer networks were supposed to replace the old-fashioned chain of command structure it appears that they haven't actually when it comes to personnel matters at the UA. So thank you all for your kindly wishes and NHM in turn wishes each of you the best of luck in the remainder of your careers. When all is said and done NHM hopes you will have achieved your objectives and will be satisfied with your accomplishments.

The momentous day, July first 2004, when NHM became the Professor Emeritus he was actually on his farm in Oregon. Sam didn't realize that he would be leaving for the farm early in the spring until it was too late for him to organize a retirement party where everyone could say goodbye, although NHM wasn't planning to disappear quite yet. When NHM returned in the fall things had changed. A new department head had been hired and was now in charge of everything. She, of course, had no idea what the local culture required when someone has retired, so Sam stepped in for her and went ahead with the plans he had formulated the previous spring. NHM was asked to submit some names of people outside of MCB he would like to see invited to a retirement luncheon, and also whether his wife and daughters would be there as well. Many of the MCB faculty came to the luncheon. So did NHM's colleagues from Math and Physics. The food, prepared by an outstanding Tucson caterer was delicious. Afterward Sam made a brief recounting of the role NHM played during his administration and showered praise upon him as the most creative scientist he knew to an extent that was indeed embarrassing. NHM is sure his family was very impressed with what Sam had to say.

NHM had forced Sam on many occasions to come down to his lab to see some incredible thing he had just discovered, and when Sam spoke about NHM he did so from the perspective of someone who really knew what was going on down there. Sam then asked for others to comment. The former Acting Head, Tom Lindell couldn't resist telling of the events that went on shortly after Sam had arrived to take over the department. Here is what he told them. When NHM was head of the department called Cellular and Developmental Biology from which MCB was formed he took advantage of a really nice old dude ranch in the Chirachua Mountains that the University had received as a gift and made available for official functions. Each year he booked the facility for a weekend research retreat. The ranch, a two to three hour drive from Tucson, provided an opportunity for hiking and exploring or just relaxing in free time between listening to talks and participating in discussions. MCB continued the traditional meetings until eventually the University sold the property and another venue had to be found. Tom had booked a weekend for MCB that by chance fell just after Sam arrived, so Sam came along and participated in the old tradition including the party usually held on the first night after dinner in the old lodge building. Tom, Sam, Marty Hewlett and NHM were assigned to the single large room in the bunkhouse. NHM was the last to get in after the traditional party was winding down. When he arrived Tom and Sam were already in bed but not yet asleep. Marty was just getting into bed. NHM told them that he usually showered at night rather than in the morning and said he hoped it wouldn't keep them up too late. No problem they assured him. NHM did not hurry in the shower, yet when he was finished none of the three of them had yet gone to sleep. NHM asked them if he should shut everything off before getting into bed. OK they said. Only NHM's bed lamp remained on. Then NHM very slowly slid into the bed. He got half way in when he became trapped by the sheets. The bed had been short-sheeted. Something's wrong NHM said, oh no. Then he shouted as loudly as he could, "God, Sam, give me a hand here, I need some help. God, I can't deflower this bed", as if it were a virgin. To which Sam replied, "welcome to camp". And everyone broke down into uncontrollable laughter. There you have it, four adults consisting of three department heads and a full professor all acting exactly like fourteen year olds! When Tom finished NHM stood to reveal the "rest of the story" for the first time. He told them, their plan had been leaked to him beforehand but he would not tell by whom. For that reason he made sure to be the last one to return to their room. He took the longest shower in history to prolong their anticipation, and he moved as slowly as he possibly could, as if he were in molasses not air, to string things out even longer. Finally he shouted as loudly as he could to make sure that others in neighboring rooms would hear it all. Sam and Tom were startled to learn the truth. Marty was out of town so he didn't take the heat.

Sam asked Mike Tabor, NHM's mathematics collaborator if he would like to make any comments. This frightened NHM because Tabor in a talk he gave to the Math Department once said, the most important thing that happened to him at Arizona was when he met NHM! NHM was in the audience and tried to become invisible. This time what Tabor said was that over the years NHM had attended more math colloquia than any of his faculty, and that his true introduction over there came when he started to ask the speakers questions. Pretty soon people in the department began to ask, who is that guy, and when he told them NHM was from Molecular Biology, they marveled at the insight he had. Indeed NHM did enjoy challenging them on their own ground. The best example that sticks in his mind was when the math speaker was a former Cavendish Professor of Physics from Cambridge, a man he knew from his affiliation with Gonville and Caius College where the speaker was a fellow. Sam Edwards' talk dealt with an analysis of the physics of sand piles. He developed some formal models and found that he became trapped by his equations into an irresolvable dilemma: from a statistical point of view there weren't sufficient variations possible to account for the physics. NHM thought of a way that might get him out of the problem, but waited afterwards until all the mathematicians had a chance to make suggestions or ask questions. Then when no one had gotten anywhere, he sprung his suggestion upon them. The main idea was to provide a means to use dynamics rather than statics to view the problem. Edwards immediately grasped the idea and said to the audience, "everything he says is true". Silence followed. Then he asked NHM, Neil how would I go about doing that? I will have to give that some thought NHM answered. Perhaps next time in Cambridge. NHM's credibility in math rose very high that day!

Then Sam came to NHM's colleague and friend John Kessler, a physics professor who had mastered the heart of physics in its entirety. Kessler had taught NHM a lot over the years, and NHM did what he could to educate Kessler in microbiology. What Kessler had to say surprised NHM. He told them that NHM used to write these poems to mark events and he never forgot a line from one of them, "forget about what you can't remember, simple things become the agenda". NHM too considered that line one of the best he had thought up. It came from a poem written for NHM's colleague Don Bourque's fiftieth birthday in which he tried to touch upon all the main features of Don's life.

Plant man
Musician
Father-and
Lover Boy

Fisherman
Toyman
Joyman-and
Sonny Boy

Never In
But always there
Move into your
Rocking Chair

Forget about
What used to be

METAMORPHISIS
At Fifty

The Bow Moves Slow
Time Begins to Show
The Back is Bent
Could have becomes CAN'T

Brace Yourself
Pace Yourself
Shift into third
Otherwise you'll look absurd

Fifty is the year
To readjust your gear
To understand the jive
Of being twice twenty five

Forget about
What you can't remember
Simple things
Become the agenda

Welcome to our CLUB
My friend
One thing more before I end

THE FIFTY CREATURE'S
MOST IMPORTANT FEATURE:
SOME THINGS STILL WORK!
FIND THEM AND ENJOY.

Don is a professor of Biochemistry and Molecular Biology who worked on plant systems. In 2006 he was on leave working at the US State Department. He was a single parent who raised his daughter (now a CalTech graduate working as an engineer), and a professional musician, a bass player in the Tucson Symphony Orchestra. Don and NHM frequently pointed out to the MCB community that our department is the only one of its kind in the world with two professional bass players in it! How do you know that, they were once challenged? To which NHM replied, we both are members of the International Society of Bassists and know many bass players from around the world. They know us too. If there were anything else like it elsewhere, we would know it. That seemed to satisfy them. Perhaps NHM should add here while he is on the subject, Don recently remarried. He asked NHM to be the best man at his wedding and NHM was happy to say yes. Several days before the event Joan told NHM that he would have to make a toast to the couple and ought to think about it beforehand. Here is what it turned out to be.

A toast is just a piece of bread
Subjected to some infrared.

No that's not it. I'll start again.

At weddings one is often said before the couple goes to bed.

Lets not get into that I was told so I had to start the third time. OK

A toast is just a wish you see
That everything will perfect be.

So let us drink to pledge our help
That this be true for the two of you.

So let us drink to pledge our help.
So let us drink to pledge.

So let us drink.

And everyone did drink. And the wedding was beautiful. Now back to NHM's retirement luncheon.

Even the new Department Head, Dr. Kathleen Dixon had a few words to say. She told us that before she arrived here she had heard that someone in the department had done this, someone had done that, someone was a musician, and someone had a farm, someone… on and on. Then she said it turned out to all be the same someone. I better meet this person she said. She connected all the dots and it turned out to be NHM. There followed the presentation of some unexpected gifts. NHM got a beautiful piece of Southwestern Indian pottery, a small basket, and of all things a plaque inscribed as follows:

MCB

TO: NEIL MENDELSON

IN APPRECIATION
FOR 35 YEARS OF DEDICATED SERVICE
AT THE UNIVERSITY OF ARIZONA

By combining Biology, Engineering and Mathematics
you have recognized problems unseen by others
and gone on to solve them

FROM THE FACULTY, STAFF AND STUDENTS

14 MAY 2004

NHM really hasn't had much experience at receiving things, at being honored, or being told how well he has done this or that and so he approached this gathering with much trepidation. To his surprise he enjoyed it immensely. He had a chance to comment on things others said and to let them know how much he appreciated their being there. It was unquestionably the high point of his career as a professor at Arizona. Afterwards he and his wife made some thank you cards using digital images from photos they had taken of their lavender farm. NHM sent them to all those who participated. Unfortunately now after having shown everybody how beautiful their place is, they have learned that they must sell the farm. And the next year they did. You will see in what is yet to come how the farm fit into the picture of NHM's life plan and why it is sad to have to give it up after owning it for only six years. Having it at all was indeed a treat, but he and his wife are now looking forward to moving on to the next adventure in their forty seven-year journey together.

11. The heart of the matter, part I

Now let's get down to the heart of the matter. How can NHM's career be understood? What are the factors that could have shaped anyone to do such things and to fight it out to the bitter end? Is it indeed true that NHM is immune to "social pressures" as claimed by Professor Weinert in the lecture he gave about NHM's work? Could it be true as NHM's late colleague Dr. Ernst Freese who spent his career at the NIH once told him that he was fortunate to get to Arizona? Ernst claimed that only in a place like that could NHM have done his own thing without the kind of pressures that he would have had to endure in an institution where he might have fit in. Until the writing of this book, NHM gave little thought to such matters. He just marched ahead as if the course he was following was somehow predetermined. He was not blind to the fact that he was being ignored or blocked at every turn, or to the fact that those opposed to anything he did outnumbered those in support by a very large factor. Somewhere along the line NHM realized that to take on the overwhelming opposition would sap his entire energies, best then to simply ignore them and play his game not theirs. What that game was and how it became established is the focus of the remainder of this story.

It is best to start with a note about NHM's parents and their parents and the culture to which they belonged. NHM's mother was born in Warsaw, the second of three children. She and her older brother came from Poland to the United States with their parents in 1918. She was five and he was nine. The family settled in New York City. A third child was born into the family in 1923. NHM's father, the eldest of three sons was born in New York City in 1909. His father's family was originally from Minsk in Belarus. Whether NHM's grandfather was born there or was the son an immigrant from there was never made clear to him. Both NHM's grandfathers were businessmen, one had been an insurance broker, the other a women's coat and suit manufacturer. NHM spent a summer as a teenager working in the Manhattan showrooms of his paternal grandfather's company but by that time his grandfather was already dead and the company was being run by his father's younger brother, the middle of the three sons. The company came to an end when NHM was in college with the murder of his uncle. It appeared that organized crime had captured all of the business assets almost immediately or so the theory went. Everything disappeared without a trace. NHM went back to take a look himself during a leave he had from the Army. At that time he was a lieutenant and he went in uniform. It was the early stages of the Vietnam War. NHM entered the building and announced to the elevator operator that he wanted to go to *Helene Juniors*, the name of the old family company, as if nothing had ever happened. He recognized the elevator operator as the same man who had been there years earlier. The operator looked at NHM without recognizing who he was and said that the firm was not there anymore. NHM said OK let me off at the loft where it used to be. Though apprehensive, he did so. The space had been completely redone. NHM could not find a single person who had worked for his grandfather's company. Later he entered the rear of the loft from the freight elevator and spoke to the laborers who were employed by the new company. Although some of them said they knew about the old company no one was willing to remember anything about it. They had no idea what happened to it or where it might have gone. The murderer of NHM's uncle was never found, nor was what happened to the company or the firm's assets. The police focused for a while on a potential suspect but nothing came of that. NHM's grandfather had built a successful business starting from nothing. He knew mechanics, was an amateur mathematician and had a good sense of fashion design. Out of that came a very successful business and wealth that passed to the murdered uncle then evaporated overnight. Perhaps the only tangible asset derived from his efforts had been NHM's father's education.

NHM's father and his siblings were raised by a stern, strict, illiterate and uneducated woman of high fashion, their mother. She may well have been the reason that NHM's grandfather became a woman's clothing designer and manufacturer. The home environment was not particularly pleasant for the children. A lot of their time was therefore spent out on the streets. Sometime before NHM's father went to high school he became friendly with a boy named Joe who lived nearby. Although the paths taken by each of them in life were very different the two remained friends throughout their entire lives. Though both of them started in the same New York City Public School they ended up in different worlds. NHM's father went on to the select high school for outstanding science students (Stuyvesant H.S.); Joe went to the local high school and worked at the same time. NHM's father went to the City College of New York and graduated summa cum laude, Joe went to work to earn a living and help support his family. When NHM's father decided that he wanted to go to medical school his college advisor reviewed his academic achievements and class standing and told him he need not apply to more than one because he would be accepted at any school he wanted to attend. He did that, and was rejected without comment. This defeat did not eliminate his enthusiasm to become a physician but he had no idea of what he should do to gain admission. A solution was found but by whom remains unknown. He would go back to Europe, to study in Vienna at their famous medical school. NHM is certain that no one in his family had ever lived there or knew anything about Vienna first hand but nevertheless this was the plan and it worked. On top of that, before he left he decided to get married and take his bride with him. The girl he chose, and who went along with the plan was none other than Joe's younger sister, Rose. Although she was only sixteen then, they married and off they went in 1930. That made Joe NHM's uncle although NHM wasn't born until seven years afterwards.

In Vienna NHM's father continued his outstanding academic performance, but his mother remained uneducated with the exception of some formal training she took in drawing and painting, areas in which she had some natural abilities. The group within which they circulated included medical students and people like Konrad Lorentz, the son of a famous orthopedic surgeon and professor who did not think much of his son's interest in animal behavior and imprinting. Of course the son later outdid his father by a great deal. He ended up with a Nobel Prize.

Whether or not the father's method to correct clubfoot is still followed NHM does not know but the son's contributions were the foundations for work still pursued today. In this setting the one most important thing NHM's mother to be did was to learn to speak German without any trace of a foreign accent. Indeed the survival of both of his parents depended upon this fortunate skill. By the time NHM's father was close to finishing his medical school education, Hitler had come to power and the brown shirts who adopted his policies were rapidly gaining control of Austria. The Jews in Austria were targets and it became uncertain whether NHM's parents to be would be able to leave the country. NHM's father could be recognized as a Jew in Vienna but not his mother. They decided as the time for departure neared that only she would go out on the streets and thus it fell upon her to make all the arrangements necessary for their leaving. She took full advantage of her language skills by creating the impression that she was a Viennese woman married to an American student. She talked her way through all the bureaucies until the necessary papers were in hand and the day that happened they immediately left the country abandoning everything they had except for a few of his medical books. It was 1936 when they made it back to the United States. Meanwhile Joe had worked himself up from a messenger on Wall Street to a stockbroker and his younger sister had grown to be thirteen. She was the first in NHM's mother's family to be born in the United States. NHM was the third. Joe's son Chuck beat him out by six months.

By the time NHM was born at the end of 1937 Europe was well on the way to the Second World War and the Nazi's were headed to their "final solution" for the Jews. NHM and his family were safely tucked away in upstate New York living on the grounds of a mental institution. Yes they lived at Binghamton State Hospital in an apartment that was attached to a ward that housed the most pathetic individuals one could imagine. NHM came to know them as friendly neighbors who rarely left their quarters although his parents had a woman named Noni who did come once a week to help his mother with household chores. NHM remembers life there well: it was a green country setting on the top of a hill that looked out over farms with cows in the meadows and people who for whatever reason needed help to live. His father was one of the doctors who helped them. That was the idea, he supposed. There were no other children to play with but NHM didn't notice that. Something interesting, at least to him, was always to be found on the porch out front or the lawn that surrounded the building. NHM and his parents sometimes went to eat in another building rather than at home. That gave him a chance to ride his tricycle along the walkways to and from home. He recalls that his father had a car with a radio in it and that once in a while they would ride in it to a place under big trees, park and listen to the spooky "Shadow" program on the radio. NHM loved it. He even loved going to a nearby lake in the summer to play at the water's edge although he nearly drowned in it the first time they went there. The biggest of all events however, even bigger than the time in winter when his cousin Chuck came to visit from New York City and it snowed, was the day his father came home with a little puppy for them. It was a Springer Spaniel that they named Spottie. Now that was something. NHM knew nothing about the war then. In fact he didn't even know they were Jewish.

The first really bad news of NHM's life came just as he was getting to know what their puppy was like. They were going to have to leave their home, give away their dog Spottie and go to live with NHM's grandparents in Brooklyn, a place he knew nothing about. The reason was that NHM's father was going to be a doctor in the army. Of course that didn't make much sense to him then but the hardest part to understand was why their dog couldn't come along with them.

Why can't we take Spottie, he asked? Because there was no way a dog could live in New York City, was the answer. So they gave their dog to the telephone operator who was very pleased to get the gift, packed their things and drove to NHM's first view of New York City. That happened in 1941 but it remains in his mind as if it were yesterday. Observer has given it a special place that can easily be visited in NHM's mind. Carroll Street in Brooklyn was the ugliest place NHM had ever seen. There was nothing green anywhere, no grass, no earth, just gray and filth. Even the air seemed dirty. They entered a large building and climbed to the first floor where NHM's grandparents greeted them. His grandmother showed them the room where NHM and his mother would sleep, his father, they said, would be away in the army working all the time in their hospital, even at night. Furniture and curtains were arranged so NHM's bed was isolated from the rest of the room that his mother used. They would all eat in the kitchen where his grandmother cooked the food. The room she and NHM's grandfather used and a bathroom were at the other end of the apartment. In the middle was a living room that you had to walk through to get to the bathroom. In it were a large couch, an easy chair, and a small end table that had a clock and a radio on it. At 6:00 p.m. each night the news came on. It was a man named Gabriel Heater who told about the war and other things. NHM's grandfather sat and listened to it. NHM had to go to bed at six so when he heard the familiar voice on the radio he went into his room and to bed. NHM's mother said goodnight and left. His grandmother worked in the kitchen nearly all the time. From little bits of news NHM heard on the radio before falling asleep he learned that lots of people were being hurt or killed in the war. It made sense then why his father would have to work day and night in the army hospital.

The now extended family quickly settled into a routine in Brooklyn that went something like this. NHM's grandmother would get up early and start in the kitchen and stay there until the night. His grandfather would get up, smoke cigarettes and read the paper or listen to the news on the radio. Occasionally he would dress in business clothes and go to work somewhere in an office. He was said to be an insurance broker but no one ever said what that meant. NHM's mother worked in the war effort as a telephone operator during the day so NHM was left to his own doings and out to the street he went. There he learned many things, perhaps most of the things he knew about anything. The other kids out there who were in the same situation as he was were his source of information. There was a bully named Karl, who tried to beat up anyone he could catch alone. After fighting with him several times NHM told his mother about him and she said she would go to speak to Karl's mother. She did and afterwards NHM asked her what was wrong with Karl. She said Karl's mother had said that he was very angry but that he wouldn't fight with NHM any more. Not long after that NHM saw a crowd of kids all fighting in front of the building where Karl lived. When he ran over he saw that they were all beating Karl to get back at him. Karl was lying on the bottom and getting pounded when his mother ran out of the house to save him. She had gotten a boat oar somewhere although there were no boats anywhere near our street, and used it to sweep waves of kids away from Karl. She dragged Karl into their building and kept him in for days afterwards. When he reappeared he never spoke to anyone again and never threatened another person. On the day the kids on the street really needed him, Karl ran home rather than help them. It was the day that a neighboring block invaded NHM's block. A whole gang of them for no reason what so ever swept through Carroll Street with sticks and chains swinging them at anyone who was out there. NHM couldn't believe it. We are all on the same side he thought, why would they do this. And he ran like hell for shelter. He ran into the basement of his building and hid behind the coal bin near the boiler.

When he thought it safe to come out he remembered that the building super, as the caretaker was called, a Mr. Polanski, had a large vicious dog tied near there to guard something, perhaps the coal? NHM somehow got by the dog on the way in but he was all teeth and raised hair on his back when NHM tried to get out. Once before NHM had gone down to the basement with his father when he visited them.They went to replace a blown fuse that the super had never gotten around to fixing. The dog tried to keep them both from entering. NHM's father simply held him at bay from the top of the rope that tied him to the boiler. NHM tried that but he wasn't tall enough to keep him at arm's length. The dog could still, bite him. It didn't work. Finally desperate to get out before getting bitten, NHM took a piece of coal, approached the dog as slowly as he could, and when the dog started snapping he threw the coal as hard as he could at the dog's head. The blow startled the dog just long enough for NHM to run past to safety. He went outside again to see if the invaders had left. There was no one on the street so he returned to his grandparents' apartment. His grandmother asked how he had gotten so dirty. He told her that he had been playing with some coal. She said, I didn't hear the truck deliver coal today. Never mind she said your mother will be angry to see all your dirty clothes when she gets home, so let me wash them up before she gets here. By doing so all the events of the day remained unknown to any of them. It turned out that was almost always the case. What happened on the streets never came to the attention of the family.

There were many reasons that NHM hated Brooklyn and the people living there but he did learn a lot of things in his daily existence on the street. He learned that the little flags with gold stars on them that hung in many windows were a sign that someone who had lived there had died in the war. He began to count the number that he could see from the street. These little flags were usually attached to the window shade so sometimes when the shade was up they couldn't be seen. He delayed coming in for dinner later and later to see how many would appear. After a while he knew all of them in his building, and when new ones appeared he came to realize that the war was still a dangerous thing. He asked his mother if his father or his father' younger brother would be killed in the war. She said she hoped not but NHM's grandmother added that she was certain that NHM's father would come back OK. NHM is not sure how he came to learn the second important thing: the reason for the war was because Hitler wanted to kill all the Jewish people. That worried him because although he had never been told they were Jewish he suspected that might be the case. No one in his family was religious so there were no rituals in the home that would have revealed that they were Jews. Several languages were spoken in the house: his grandfather spoke mostly Yiddish so most of what he said NHM never understood. NHM's grandmother always told him to "speak English to the boy". He did but when anything important had to be said he always switched back to Yiddish. NHM's grandmother, whose sister lived not far away in Brooklyn, always spoke Polish to her. And NHM's mother would on occasion speak German. Once in a while his grandfather also spoke French. NHM had no idea what these languages were at the time. Only as an adult did he piece together what they were. But on the street he learned that only Jewish people spoke Yiddish. NHM didn't have the nerve to ask at home whether that was true or if in fact they were Jewish. On the street they said you must be Jewish but NHM wasn't really sure. He had no idea of whether any of the others on the street might be Jewish or not but he was worried that if he was the only one it might be even more dangerous than he had come to expect in New York, or anywhere for that matter. So he decided to just keep a low profile and to try to figure it out on a day to day basis.

The emptiness of Carroll Street is hard to forget. There were virtually no men to be seen. NHM remembers only three in their building, his grandfather, the super, and the husband of his mother's friend Ethel. In the neighboring building observer can recall only the super and a man who served as the air-raid warden. Periodically when the sirens sounded, the warden appeared in the street wearing his armband and cleared everyone into shelters or buildings. All lights had to be turned off or blocked from view from outside. If any could be seen he ordered them to be put out. NHM often got caught out on the street in such drills and couldn't get back to his building. At one end of Carroll Street there was a footbridge over some railroad tracks that led to another street where his grandfather bought his newspapers and cigarettes at a small store. His grandfather often sent NHM to get them. One time when he was there, the siren sounded. NHM thought he might be able to run home quickly enough to avoid the warden but he was wrong. As soon as he crossed the bridge back to his street the warden told him to run back to the store. There, the shopkeeper and NHM went to the basement and awaited the all clear signal. When NHM got home his grandfather asked what took so long. NHM explained that the warden had sent him back to the store. His grandfather became angry and said the warden was a dunce. NHM's grandmother said that even if he were I would still have to obey his orders. That was more important than getting the cigarettes back quickly, she said.

Although Carroll Street was indeed a true street neither cars nor trucks seemed to use it. The only car NHM ever saw there was his father's the day they arrived and when he brought NHM's infant sister and mother home from the hospital where his sister was born. The few times his father came back to visit during the five years NHM lived there he must not have come by car for it never again appeared on the street. Every so often the coal truck arrived. You couldn't miss it. The noise of the coal going down the chute into the basement drew all the kids out immediately. They all mulled around the chute hoping to get a piece of coal that bounced out of it. Coal could be used like chalk to write on the street. Another large truck also came routinely to remove the cans of ash produced from burning coal and garbage. There were incinerators in each building that the super used to burn the garbage. All the garbage went down a dumbwaiter from NHM's apartment each day. The super burned it whenever he had a big enough load accumulated. The buildings were only five stories high and thus the incinerator chimney was low enough so that everyone could always smell the burning stench. At night sparks could be seen blowing out of the chimney. Kids on the street sometimes tried to catch a live cinder but they usually burned out before they wafted down to the street. One day a third vehicle appeared on Carroll Street that had not been seen before. It was some kind of a service truck. Two workers were in it. They had a variety of tools and machines that drew much attention. Once they got set up they dug a hole in the street to reach whatever it was that they had to service. NHM was blown away by what they revealed. Under the street was earth. He had no idea such was the case. NHM had assumed that New York was built on a rock or something like that. When he realized it was earth he recalls thinking that people were responsible for ruining this place, not nature. People must have covered it all over. Why they should have done this escaped him.

The only places NHM ever went from Carroll Street during the war years besides the newspaper/cigarette store were a few outings he had with his grandmother. They went several times to her sister's home about a twenty-minute walk from their apartment, but a world away in everything else. Her sister's home was a brownstone building in which her sister's husband had a dental office in addition to their living space. The house was beautiful and well furnished. Although they too were refugees from Eastern Europe, they must have been much richer than NHM's grandparents were. The dentist was also an accomplished musician. There was a piano in a music room and several mandolins. NHM was allowed to touch several things and to try the piano. He was immediately hooked. In addition he noticed that the children living there, his second cousins, had toys. The eldest boy had a bicycle. Only one child on Carroll Street had a bicycle. Sometimes he let others ride on it. NHM taught himself how to ride a two-wheeler on one such occasion.

Indeed things seemed to be much better in the brownstone on Ocean Parkway than they were in the apartment on Carroll Street. NHM tried to memorize how to walk from where he lived to their house in case he had to go there alone in an air raid or something but it wasn't easy to get the route. They were the only relatives he knew who lived anywhere nearby.

In a slightly different direction, NHM's grandmother took him one day to the Brooklyn Museum of Art. It was his first visit to a museum and it was an eye opener. The objects themselves were marvelous but the real serious part was the information they conveyed about life. Some paintings reminded him of life at the mental hospital, something he had nearly forgotten. Others revealed things beyond his dreams. In this building NHM found stories that somehow did not penetrate to Carroll Street, or at least to his part of Carroll Street. That visit was one of the most exciting days for Observer and Creator too. It was too bad that NHM's grandmother could only occasionally find the time to go on such outings and he was too small to go there by himself. NHM was effectively trapped on Carroll Street where information was hard to come by. Mostly he heard only about the war but it wasn't all useless information. One day he and his grandmother went to the Brooklyn Botanical Gardens. His grandmother sat down on a bench to rest while he wandered about looking at the flowering trees and plants. Then he heard a kind of grinding sound but couldn't determine where it came from. When he located it in the sky it appeared to be an enormous bomb coming toward them. It was very frightening. The "bomb" seemed to be going much more slowly than he had imagined it would travel, and it appeared not to be coming down but rather moving horizontally. This didn't make sense but it looked like a bomb he had seen on the cover of one of his grandfather's newspapers. So he ran to his grandmother to alert her, and asked her if it was a bomb. No she said it's a dirigible. What is that he asked? It's a kind of airplane she said that flies very low above the ground. The army uses them to find spies who send messages to the Germans about New York. If we catch them it could save a lot of lives, she said. How do they send the messages NHM asked? They use a kind of radio she explained that they talk into and it goes through the air just like the radio we listen to at home. The Germans have radios listening for messages the spies send them. NHM watched quietly as the big ship passed over them and went off on its flight. The hum of the engines was very loud when it was just overhead. No harm came of it. But NHM considered it a close call. What if it had been a bomb he thought. It could have killed anyone near where it landed. On the way back to Carroll Street NHM wondered how he might find his father if a bomb killed everyone in their apartment when he was out in the street. He asked his mother if she could teach him how to use the phone since she was an operator. She said she would when he got bigger but that if he had to use one before that he should go to Ethel's apartment and ask her to help That sounded reasonable. NHM knew where Ethel lived.

One day as summer approached when NHM was about six, his mother told him that they were going to live in the country for a little while so they could be closer to where his father was now working in an army hospital. Do you mean we are going to move away from Carroll Street, he asked? Yes she said but we will have to come back at the end of the summer. So the two of them left for their summer vacation in New Jersey. In the country they lived in their own apartment in an old farmhouse. NHM's father came home every night to stay with them. Nearby a boy had a pony but he wouldn't let NHM on it. He said his mother had to put him on and if he got off he couldn't get back on. So NHM sometimes walked along with him as he rode around and showed NHM things in the country. NHM loved it. He wished he could have stayed forever. But everything wasn't perfect. NHM had a small accident. He slipped on a small throw rug atop a wood floor and on the way down his arm went through a glass pane of an internal door. The glass cut his arm to the bone, but did so very neatly, like a fine slice. There was no blood and it didn't hurt but you could see the bone and it looked frightening. NHM showed his mother and said I guess we will have to go to the hospital. She got very shaky and told him to call an ambulance. He didn't know exactly how to do that, so he dialed the operator and told her the problem. She said there was no ambulance available but he should go to a doctor's office. Where NHM asked? The operator asked where he lived, and he asked his mother. He relayed what she told him to her. The operator figured out where they should go. His mother called a cab to take them there. NHM wrapped a kitchen towel around the cut. His mother said she did not feel well but she came along with him to the doctor's office. When the doctor saw how wide the opening was he ruled out closing it with stitches. Instead he said he had some wide surgical staples. They might hurt he said. But they really didn't. Afterwards the cab took them home. NHM found the most disturbing part of the whole incident the fact that his mother could not function under pressure. He realized that back on Carroll Street his grandmother made all the decisions about everything. His mother in some ways behaved like a child herself who had gone back to live with her parents. NHM realized that it was primarily his grandmother who was raising him on Carroll Street.

With the exception of the one-day that his cousin Chuck and Chuck's mother came to visit, no other relatives ever came to see them in Brooklyn. NHM never saw his paternal grandparents who lived somewhere else in New York City or other relatives on his mother's side of the family. His mother's older brother Joe and her younger sister also lived somewhere in New York but wherever it was they never came to the Carroll Street apartment. NHM heard that his mother's younger sister was a student in nursing school. She had gotten the idea to become a nurse when his father went away to medical school with her sister. When she became old enough to enter nursing school her father refused to provide the ten dollar fee she needed to enroll. Grandpa did not believe that women should be educated although his wife, NHM's grandmother, had studied in Poland (he himself went to Paris as a student). Since no one had any funds independent of Grandpa in the Carroll Street household, NHM's aunt was up against a barrier that seemed insurmountable. It would have all failed had it not been for Joe. Joe had a job on Wall Street and was able to save up the ten bucks his sister needed. Joe's entire life was filled with deeds like that. He found a way that got them all out of Brooklyn at the end of the war. When VE-day came and the Germans had been defeated there was great excitement on the streets. Aunt May came from Nursing School and together with NHM went out to bang on pots in the street along with everyone else on Carroll Street. That was the day NHM learned they were Jewish. His aunt told him. But she was careful to add that the dentist who was married to his grandmother's sister was not, and neither was their son. NHM asked his mother that night whether his father was going to return and if they could then move to their own place. She said yes they would move soon but that his father couldn't come back yet because there were going to be a lot of soldiers returning to America after fighting who would need his help. She told NHM that his father had become a specialized doctor, a psychiatrist and neurologist who could help people who had become sick while fighting. She said there weren't very many doctors who knew what he did so he would have to stay in the army for another year. But we were going to move before that. That sounded great to NHM. His younger sister had been born in 1945 and the room he shared with his mother was partitioned yet again to create a place for her crib, along with his bed and his mother's space. It was hard to sleep there when his baby sister cried at night. NHM couldn't wait to leave Brooklyn. There was very little joy there.

About two years before the end of the war NHM's uncle Joe learned from a client of his that the government was trying to plan for the return of soldiers when the fighting was finally over. There was going to be a very severe housing shortage so if they ever hoped to move from Brooklyn they would have to plan way ahead. Joe had heard about a very large development being built by the Metropolitan Life Insurance Company in the Bronx called Park Chester. It was a new idea. There would be a planned group of apartment buildings with open green space and some shops in a kind of self-contained village. It sounded great. When Joe learned that the company had begun taking reservations for apartments he immediately signed up for three of them, one for NHM's family, one for NHM's grandparents, and one for himself and his family. It wasn't clear then that they would get all of them but when the time came, they did. NHM's father came to help his mother and the two children move into their new apartment. They really didn't have any furniture except for his sister's crib, so the bare essentials were bought. The two bedroom unit in Park Chester was very sparsely furnished with what they had. It was on the sixth floor of an eleven-story building. From the window people on the street looked liked large ants. As soon as NHM got acclimated he realized that their new place was much inferior to the Brooklyn apartment. Being inside these new buildings was more or less like being inside a prison. The buildings were all rectangular with long dark halls on each floor. There were no windows in these corridors, only the same locked and double locked doors, one after another along convergent lines, as any illusion of depth would create. Not a sound could be heard, nor a person seen. If a door opened suddenly the breaking of silence caused a rush of adrenaline. Who knew who or what might come out? NHM certainly didn't. He will never forget the day he retuned home from a day camp his mother had sent him to for part of the summer. A large car picked children up in the morning drove them to the camp, which was along the water under the Whitestone Bridge, and returned them home at 5:00 p.m.. When he got out of the elevator on six, NHM had to go to the bathroom so he rushed to their door. He knocked but no one answered. He knocked again. No response. His mother wasn't home. Where could she be? She must have gone out with his sister but to where? He had no idea. The only neighbor he knew lived across the hall from them. Although hesitant he knocked on their door but they too were gone. So he decided he better go to his grandparents' apartment at the other end of Park Chester. He had no idea how he could get in touch with his father or where he worked then. So he set out in the direction of his grandparents place walking very quickly so he could get to the bathroom. About halfway there he met his mother strolling along with his sister in her carriage and another woman with her child in a carriage too. When his mother saw him she said, oh is it 5:00 p.m. already? NHM said he thought it was. And together they strolled back home. Later he asked her if he could get a key. She said she would ask his father. NHM wondered how she would. He didn't live there with them. NHM never heard his response. Camp ended but nothing changed. No, he never did get a key to their Park Chester apartment. He was nine years old then. His mother appeared to be totally occupied with his baby sister's needs. He believed she assumed NHM would be perfectly occupied on the street as he had been in Brooklyn. The main difference, however, was that in Park Chester his grandmother wasn't around any more to answer questions.

NHM went to two schools during his years in Park Chester. The Public School was several blocks outside of the complex. He does not remember having been taught a single thing there. NHM had never learned to read in Brooklyn and was therefore behind all the others. There were few books available to even look at. His mother had sometimes read Uncle Wiggly stories to him when he was going to bed in Brooklyn but he couldn't see the pages in the dim room. In fact he couldn't see very well at all. Eventually it was recognized that he needed glasses and he had gotten them just before the move to the Bronx. Although the eyeglasses enabled him to see the words it didn't convey what they meant! He hunted for a book at home or even a newspaper to search for words he thought he knew. He did eventually find a book he had gotten as a gift in 1944 from his mother but knew nothing about it. It was a Children's Bible, both the old and new testaments. NHM still has it. It was inscribed to him as follows:

Neil H. Mendelson, State Hospital, Binghamton

Next page: Neil H. Mendelson, 916 Carroll Street, Brooklyn, NY

To my son on his seventh birthday

War Year 1944

(It must have been purchased before they moved to Brooklyn in 1941, and then given to him as a gift afterwards.) Two years after having gotten the gift he could barely read the inscription. He searched in the book for words. There were few that he recognized. For some reason NHM was never taught to read. It appeared to him that everyone else knew how by age nine.

The second school he went to in Park Chester was called the Park Chester School of Music. It was located in a small old house just outside of the complex. NHM was sent there because his grandmother had told his mother that on their visits to her sister's, NHM seemed to like the piano. Indeed he did but unfortunately the school only had a single piano that was reserved for practice time rather than lessons. In class the student's were given wooden boards that had the keys painted on them. The instructor had them place their fingers on the board and press the board one finger at a time as she told them which note would be sounded. Then the students got assignments to work on at home after class. NHM didn't have a piano at home so he went back to the school on some afternoons when their piano was available. On the real thing, he played tunes by ear. He made a startling discovery. The same tune could be played starting on any note. It took some years for him to realize what that really meant. All the while NHM's father infrequently visited them. He appeared not to know anything about NHM's schools. Rather he focused on whether they were all well and what the plans might be for the time after he would leave the army.

A trolley car ran on a street, Tremont Avenue, one block beyond the edge of the Park Chester complex. It went directly to a terminus near the Bronx Zoo. The Zoo was the only interesting place that NHM could get to but he wasn't allowed to go alone. He worked on finding others who might be able to go with one of their parents. His own parents were so occupied with other things it was hard for him to get them to go. In the trolley NHM often sat at the opposite end in the conductor's seat used when the trolley was traveling in the opposite direction. All the kids did that making believe they were driving the trolley. Someone on the street told NHM that they had gotten a new animal at the Zoo never before seen in America, the Duck-Billed Platypus, whatever that was. NHM told his mother they had to go see it because it wasn't certain that it could live in a Zoo. She believed that and agreed that they would make a trip there. There was of course a long line of people waiting to get a look at it. When they got close it wasn't that easy to see it but even so you could get a good enough idea of what it was like. NHM's mother wasn't impressed but she went along with it. On the way home NHM drove the trolley backwards as usual. That night he had a strange dream. It was more about the trolley than the platypus. He dreamed that his family and grandparents were all in the trolley but no one else and that he was driving it. Then in some way he was able to fly it up into the sky off the tracks. He flew it to a great place somewhere where they landed. And they stayed there. It was kind of a magic escape. And it was a recurrent dream. Events did not work out exactly that way.

NHM had at least two other places to live before he could even get away himself; the first of these two was even more undesirable than either of his first two homes in New York City.

When NHM's father did finally leave the army he established a private practice in Manhattan and became affiliated with several hospitals located close to it. Although he was board certified in two specialties he sought more qualifications, first as an analyst and then later as a Mental Hospital Administrator. To qualify for the former he had to undergo analysis himself. That was not a pleasant experience for him or for NHM's family. Observer will get to that but first came their move from Park Chester. NHM's father decided that they should purchase a small attached-house. Houses were being constructed all over the city to provide space for veterans many of whom were living in temporary quonset hut developments that were made available to them. In between all his other obligations NHM's father searched for and found a group of homes being constructed near LaGuardia Airport in Queens. He took them there to see it. The units consisted of a basement, main floor and upstairs with three bedrooms and a bathroom. Out front there was a small lawn, and in the back there was a yard about as big as the house itself that ended on a service road that separated one row of houses from the backs of the neighboring row. The newly found place didn't exactly match the kind of place they were flying to in NHM's dreams but it did have a private room for him and another for his sister.That seemed like true luxury. NHM's father appeared to like it well enough to call his youngest brother, who had survived battle in the Pacific and was living at the time in one of the quonset huts, and to suggest to him that they both buy one. With financial backing from their father they did so although the two units turned out not adjacent to one another. Units were selling so quickly that after NHM's father committed to buy one, the neighboring nine units were sold before his uncle signed his contract to purchase. So the two brothers ended up living ten units away from one another. On NHM's first night in the new house the error of his father's choice became apparent. Across the street in the nearby airport hangers aircraft engines were mounted on blocks and ran all night as part of the required routine service and upkeep. The noise was deafening but there was no way out. NHM's family was trapped there as was his uncle's family down the block. When NHM explored the neighborhood over the next week or so he came upon horror number two. At the end of the block on which the house was located the street ended at Flushing Bay. At first NHM thought that this might be a great place to see fish, birds, and other shore things but he soon discovered there was a terrible odor near the water. At first he thought it might be from the aircraft fuel that was downloaded from barges there into huge storage tanks. But it wasn't. No it came from a flow underneath eighty first street that emptied into the Bay. It was raw sewage. NHM told his mother but she had no comment. He wasn't certain whether she had told his father so NHM repeated it to him the next day. He too had no response. Neither of them would walk down the street with him to see the problem. From time to time NHM returned to see whether the problem persisted. Whenever he did he could see feces floating out into the Bay and it stank. Still they continued to live there as if the open sewer never existed.

NHM's father went to work everyday before NHM had awakened and returned long after the family had eaten dinner. He was always in a foul mood. On weekends he insisted that NHM work with him on projects around the house, which NHM did regardless of whether he had made any other previous plans. NHM asked his father if he could help him with some projects for school but he never really had the time to do so. NHM's mother said it would be better not to ask him again until he first sought help from her. Needless to say that with the deficits NHM brought with him from Brooklyn and Park Chester he fell further and further behind in the new school. When he finished the sixth grade he had to move to a junior high school a bus ride away in Astoria. The few friends he had from his neighborhood all were placed in the advanced class that compressed three years into two before going on to high school. NHM was placed in the fifth class down, grouped with others who knew nothing. There he learned nothing except about the Mafia. There was a hierarchy in the junior high school in which each student was given a position on the basis of his ruthlessness or power. Those at the top of the ladder enforced all the rules as a gang. To show power they built and brought weapons to school. One day they took what looked like a canon from the trunk of a parked car near the school and fired it across the schoolyard through a series of ash-filled cans all lined up to show how the projectile would penetrate them. In the shop class they built what were called zip guns, pistols that could fire a low caliber bullet. NHM saw the shop teacher one-day show to another teacher a beautifully carved wooden pistol grip that one of the students had made. The work was a sign of real talent wasted in the criminal setting in which they lived. After showing it, the shop teacher broke it in half. The teachers themselves were not of the academic type. One of them was definitely a patient for NHM's father. He spoke only in a whisper and naturally the hoodlum students shouted him down every day. One day he took some spoons out of his desk drawer and tapped them together to make a ringing sound. He moved the singing spoons closer and farther away from his ears and listened intensely to them. The students sang with them in swells and silences then broke into laughter. This continued day after day until the principal, two teachers and two attendants appeared to speak with him. He gathered his spoons and left with them presumably on a trip to the mental hospital. The class went wild in cheering. The chaos in the classroom was hard to curtail. NHM told his mother all about it that evening. But it took more than that to trigger moving away.

There were two events that NHM believed set the stage for their leaving. The first happened on the forth of July. Not far from his uncle's house down the block towards the Bay lived a Mafia criminal known to everyone in the community. NHM knew his son and had been in their house on several occasions. It was always dark in there closed in by shades and curtains. NHM was told by his son not to speak. The father either sat watching a tiny television, one of the first ever made, or around a table with men who worked for him. He had gotten some large aerial fireworks that he launched from his backyard on the fourth of July. Everyone went out to watch the display. Some of the ash drifted down into our yard and a piece got into NHM's mother's eye. After his father washed it out they came back out to tell NHM not to look up at the sky. A police car then drove down the alley and stopped at their yard. The officer asked if they had seen fireworks and if they knew where they had come from. NHM's mother said yes we had and that a piece of ash had fallen into her eye. Who shot them off he asked? She said someone further down the alley. The officer then said, we must shut them down as quickly as possible. Several of their aerial displays had come down on the fuel storage tanks in the airport and had triggered a fire detection system. No fuel could be withdrawn until the hazard was removed and off he went to find the culprits. NHM asked his father if it could have set fire to the tanks. He said yes it was very dangerous. That was number one. Number two followed a few months later. An aircraft that had taken off from LaGuardia had developed mechanical troubles and attempted to get back to land by flying over NHM's house. He was on the street and saw it coming. It was a twin-engine freight transport and the pilot was struggling to keep it upright. It cleared NHM's house by perhaps ten feet. He could tell that because at the other end of his street there was a field that he used to play in that was on the airport's landing approach. When the planes came in for a landing the wheels were close enough to be hit by throwing up a broomstick. Some of the kids thought that was a great sport. The plane cleared NHM's house but the row of houses behind his was one story taller and the plane scraped the high point on one of those roofs. From there the plane continued out over the bay trying to circle back to the runway that jutted out into the bay but it couldn't sweep a wide enough arc to line up properly. It came in almost perpendicular to the runway and went right off the other side down towards the water but stopped short before getting fully into the bay. NHM could see the pilots being taken out to a waiting ambulance. Several neighbors watched the whole thing along with him. One of them said he believed he saw a piece of the plane drop onto the roof of NHM's house Another said, no it was the roof across the alley. NHM told that to his mother. She said they would have to wait until his father got home for him to go up and look. NHM wondered if there might be a hot piece of metal on their roof that could set fire to it, so he asked a neighbor if he could look for them. He did. There is nothing on your roof he said but there is something he thought might be part of the plane on the higher roof across the alley. When he came down from the roof he called the airport and they sent over some men to retrieve it. That night NHM told it all to his father. He insisted on going up to look at the roof himself even though it was already dark.When he climbed down he said he couldn't find any parts of the airplane or damage to our roof. NHM wondered if he could have seen anything up there at all.

Towards the end of NHM's year in the Astoria junior high school his father announced that they would have to move again and the search, this time for single family rather than an attached house, began. By the time they made their next move NHM had settled upon a plan for himself. He decided that the best thing he could do in life was to become a farmer. It would be a definite way to get out of the city and away from the awful people who lived there. He made the assumption that farmers usually worked alone and they had to be out in the country to grow their things. NHM didn't know anybody who was or had been a farmer to speak to about this, nor did he have a route figured out as to how you could get there from where he was starting but he set about the planning nevertheless. NHM decided not to tell anyone in his family about this because he felt it would be too difficult to explain to them what drove him to want such a career and he was certain they wouldn't be able to give any helpful advice. He also felt that he was so far behind in his education that he would never be able to catch up and be competitive for the kind of career his parents would assume he would seek. NHM's mother always told him how good it would be to be a doctor like his father but he was very skeptical about that. Once the parents of a friend, both of whom were teachers, asked him if he was going to be a physician like his father. When he said definitely not they were rather taken aback. Why not the father asked? Because I don't believe you have any life of your own once you are a physician, NHM replied. My father can do nothing but respond to patients needs. He has no other life. NHM thinks they were shocked to hear that from a ten-year-old. NHM's choice to avoid medicine at all costs grew much stronger during the five years they lived in their new home, itself a beautiful mansion but devoid of life. During that period his father became more and more hostile and sullen. He would frequently explode in anger into an uncontrollable rage over trivial things. NHM's mother was often terrified of him. To avoid confrontation she retreated to a chain-smoking do-nothing individual who sat all day reading novels.

NHM was not so intimidated, consequently his father threatened him many times. He threw things at him, chased him with an axe, and tried to beat him but NHM had become bigger than his father was at that time and would restrain him. NHM found he could not rely upon his father for accurate information. On the second day of high school NHM met a friend of his who had gone to the same eighth grade school that he had, the son of a very successful surgeon. He too had been through some hard times when his parents divorced and he and his sister went to live with his very poor mother, a fine artist. The first thing he said to NHM was, "Neil, I missed you in Latin class yesterday, what happened?" "I'm taking German not Latin," NHM said, both my parents know German. Wait don't you know you can't get into college these days without Latin he said, my father told me that. Really, NHM asked him? Yes, he said ask your father. That afternoon NHM asked his mother if it would be OK for him to ask his father about it when he got home without triggering his anger. She said she thought it would be all right. So later when he was working in the office he had in their large house she knocked on his door and told him that NHM wanted to ask him a question about school. When NHM went in his mother stayed there. NHM told his father what he had heard and his mother asked him if it was true. Yes it is he said, I studied Latin for five years. The next day NHM dropped German and switched into Latin. It was two years of torture for nothing. There was no such requirement in any college NHM ever heard of. In fact later in life he had to do both scientific German and French as part of the qualification for his Ph.D.!

The day that he entered high school NHM knew that he knew nothing. His hope was that he could learn something there. Start from scratch so to speak. He found this worked very well in subjects such as mathematics and science but much less so in things like languages, history, civics and so on where either prior knowledge or life experiences were an important foundation. Survival skills on the street or at home meant nothing now. NHM had to learn to use a library quickly and to budget his time wisely. He wanted to do well enough in high school so he could go to college and college would be the place he could learn to be a farmer. It fit the plan quite well but still he didn't reveal it to anyone. To achieve all this NHM tended to his schoolwork assignments before all else. He had no help in it but he pushed ahead as best he could. In fact he learned during that phase how to be totally independent. And he was not ashamed of what he could accomplish by himself. Finally there was a chance he might learn something! Math was wonderful. Biology was too. Physics was the best. And music was the joy of his life. They had gotten a piano at home before they moved from the place across the street from the airport but there were no suitable teachers available so NHM tried to teach himself how to play. He learned the popular songs of the 1940's and 50's but he couldn't read music. He knew the words of most of the songs, however, and they too were windows to life on the outside for him, the outside of his bleak existence. NHM was fascinated by them and loved them. Indeed sixty years later he still plays and enjoys them. Listening to him play these songs his mother thought he knew how to play the piano. He knew better. Yet she insisted that he didn't need a teacher. So NHM can add playing the piano to all the other things he never was taught.

Somehow NHM had learned to play the bugle. Starting at a camp, then in the Boy Scouts he mastered it. By the time he got to high school he thought he should move up to the trumpet. He went for an audition. The orchestra teacher asked if he had ever played the trumpet. NHM told her that he hadn't but that he could play the bugle. She gave him a fingering chart for an octave scale and asked him to try it on a school instrument. It was a cinch and the horn sounded beautiful. That did it. NHM was admitted to the orchestra beginner's class and allowed to borrow the trumpet for the school year. Over the next six months he pushed himself up in the class high enough to get a substitute place in the full orchestra, a fifty-member group of real players. By the end of the second year he was playing in many school concerts. He dreamed of playing in jazz bands and at dances but managed only once or twice to get a chance to do so at some school-related parties. Then at the end of the year NHM got really sick with debilitating asthma. There weren't any very effective medications then but he managed to keep going to school although he had to stop playing sports and the trumpet or bugle. He returned the trumpet to school and gave the bugle to one of his best friends who always liked to fool around with it. Then he went back to the orchestra teacher and asked if there was any other instrument he might try to play so he could get back into the orchestra before he graduated. She asked if NHM would like to try the double bass. The orchestra was always short of bass players she said, if it goes well there is a chance that you could get back in.

Obviously NHM didn't have a bass at home nor could he get one quickly so the teacher arranged that he could practice with a school bass in a private room to which the teacher gave him a key. NHM worked at it as best he could given his course schedule. He liked the bass, though not as much as the trumpet, and he wasn't very good at playing it. NHM decided he would continue with it even if he didn't get into the orchestra. No decision about an orchestra position was reached by the end of the term, so in order to keep playing over the summer NHM withdrew one hundred dollars from the money he had earned working in a neighborhood luncheonette and went to Manhattan in search of a bass. He went from place to place asking if there was anything they had in the range that he could afford. The days search netted two possibilities, one a wonderful sounding old German bass badly in need of repair. You could see light coming through it all over the belly and back. The other was a somewhat worn sturdy plywood bass that had a nice but not loud sound. Though NHM wanted the former he settled for the latter in the belief that it would better survive the beating it was likely to take in his hands. For the hundred dollars the seller threw in a cheap French bow and an old canvas bass case. NHM dragged it all home on the subway and the bus. He kept it in his room and played it only when his father was not home. In the beginning of his senior year NHM was offered a place in the school orchestra's bass section. He was definitely the weakest of the three bass players but he was happy to be there. He wished he could have had a private teacher but that wasn't possible so his progress was very slow.

About half way through the year the orchestra teacher and music director surprised him by asking if he would like to audition for the position of student conductor. NHM was flattered and stunned when he later won the position. He got to conduct several concerts and even Elgar's "Pomp and Circumstance" at his own graduation! Over a thousand students marched down the aisle to it and there he was in cap and gown conducting. On top of that he unexpectedly received the music award given to either a member of the orchestra or chorus upon graduation. As soon as he received the award the orchestra teacher signaled for him to come over to her. She said the next thing on the program; some songs to be sung by the best soprano in the school's choir would be cancelled. NHM was asked to go back out on stage to announce that we were sorry she wouldn't be able to sing for us, which he did. On the way back to the orchestra pit he realized that the singer must have been a candidate for the music award that he received and was upset that she hadn't also been honored. NHM knew she was a much more highly trained performer than he was and probably much more talented than he was. He knew his foundation for this award was rather shaky, and he wished that somehow he could have shared the prize with her. It was a very early lesson in how one person's good fortune was another's bad fortune. There were two witnesses to all of this from NHM's family, his mother and her younger sister, the aunt who on VE Day had informed him that he was Jewish. NHM had not expected his father to attend and he didn't. He assumed his father's focus on homicidal, suicidal or other unfortunates would obviously take priority. This high school stuff by comparison was trivial. The guilt of winning the music award without really knowing how to play his instrument very well stuck with NHM until finally at nearly age forty he managed to find a teacher who taught him how to play the bass. Thank God he did. It permitted him to perform classical music in a symphony orchestra for over twenty nine years. More later.

Although they knew he liked music, no one in NHM's family knew or asked what he planned to study in college but he did tell his biology teacher at high school. The teacher was surprised to say the least, probably shocked. When asthma first struck, NHM had to stay indoors for many months in the spring, summer and fall. His father gave him a small child's microscope. It was about one half the size of a true microscope but it had good optics and NHM became proficient at using it. His mother discovered one day that NHM was growing all kinds of possibly dangerous things in his room such as mold on milk soaked bread and the like. She wanted him to throw it all out but he resisted. His father heard of all this and asked if NHM had considered examining things from the fish tanks he had in his room. He told his father that he had but was unable to identify the things that he found. We have no appropriate books NHM explained. To solve this problem his father made the following suggestion. He said he had a collection of microscope slides that he had made while in medical school that he would let NHM study, and that he had a book that would help explain what they showed. NHM said that he would like that and so his father gave him the slide collection made some twenty years earlier. NHM worked with them for some time before the promised book could be found. When it did appear he dove into it to get some answers. Indeed he could match up what he saw in his microscope with figures in the book but unfortunately it was a German text and NHM couldn't learn a thing from it. Basically what he learned from his own observations was how cells took different forms in different tissues, what their substructure looked like, and how the cells were organized in different tissues. When the topic of cells was studied in the biology class at high school NHM tried to get all his questions answered. It must have appeared to the teacher that he knew a lot about cytology and histology of human cells and NHM assumed the teacher thought he would be headed to a pre-med major in college, not to farming. NHM also confided to this teacher that he hoped to get as far away from New York City as he could in college. The teacher advised that NHM shouldn't make any final decisions at the moment since there were still two years left before he finished high school. Further, he suggested that NHM start reading about different colleges and places, to get familiar with the kinds of things they taught and what it was like to be a student at them. How do I do that NHM asked him? Well why not start by visiting one of the high school's "college advisors" he suggested. Thanks NHM said, I will. That was the first NHM ever heard about such people. He went to see one and learned from him that there were many books published each year that gave information about almost every college, what they taught, and details about such things as how big their libraries were, how many students attended them and what their admission requirements were. What are your educational goals he asked? NHM said he wasn't yet sure but that he definitely wanted to go to college. Well, you'll have to get some good grades he said. I'll try NHM told him.

In his high school library NHM found some of the books the college advisor had been talking about. He learned that you could purchase them at bookstores for a reasonable price, that is an amount that he could afford with his own money. Before he purchased any however he wrote down as much information as he could from the library books about colleges that met his goals. He noted that his grades were not good enough to get into the best rated schools but that there were quite a few places that taught agriculture that would probably accept him. The three schools that attracted him the most were a college in Colorado, another in the state of Washington, and one in Minnesota. Of course NHM didn't know anything about these places, the kinds of people who lived there, or even if he could live there, but that all seemed secondary. They all met his primary requirements. He would be able to study agriculture at any of them, they were very far away from the problems that he wished to escape, and they were likely to let him in if their published information was to be believed. At the time NHM was gathering this vital information many of his friends were busy getting ideas about what they wanted to become and where they would like to go to college. Many of them wanted to become engineers and physicians. Some just wanted to go to certain schools hoping to decide later when they were in college eventually what they wanted to do. No one ever spoke of going to the kinds of colleges that NHM was considering. Even students who had poorer grades than he did aspired to go to prestigious schools. NHM wondered how they would get in? In one case that struck close to home he got the answer. His close friend, the one to whom he had given his bugle, the same one who had told him that his father said we had to take Latin, told NHM that his father had decided that he would transfer to another school for his last two years. Transfer to where, and why? NHM asked him. It's a private prep school, he said, the same one my father had gone to, and it will help get me into a really good college, he replied. Have you decided what you would like to be NHM asked him? Yes he said I want to be an opera singer but my father wants me to be a doctor as he is. NHM wasn't surprised by his answer. On Christmas Eve the year before the two of them went through their neighborhood. They stopped at each house, rang the doorbell and NHM's friend began by singing Silent

Night, then NHM played it on his trumpet, then they ended by both performing it together. For this they usually were given a buck, which they split firty-fifty. Laddie had a beautiful voice indeed, and had some training as well from the music director of his church choir. Well I hope it works out NHM told him. (He ended up an ophthalmologist!)

Slowly but surely NHM's own victory day was approaching, that is the day that he would depart for college and start the rest of his life but first he had to make applications and get accepted somewhere. He went to college day at the high school but there was nothing relevant there for him. He also went to see one of the college advisors to find out what he had to do to submit applications. The first thing he told NHM was that he was only allowed to submit three applications. There was an application fee for each one he said, so NHM knew he would have to finally let his parents know what his plans were. He began that evening by telling his mother that he would need money to submit college applications and the sooner he could get them done the better because if he didn't get in to any he would have to try at other schools. She relayed the information to his father who asked NHM for more details the following evening. Where do you want to go and what do you want to study he asked. NHM told him. Well he said, if you want to study agriculture you will have to apply to Cornell University and to Rutgers University, the agriculture schools in New York and New Jersey. My grades aren't good enough to get into those schools NHM told him, so I would like to apply to my three choices. Well you apply to Cornell and Rutgers first, then one of the others he said. I once was on the Cornell campus. It's very nice there. You will like it. OK, NHM said. The next day he went back to the college advisor and told him his father wanted him to apply to five not three colleges so what should he do. The advisor said to fill in the applications and if they have time for the extra two they would process them but he couldn't be sure because they already had more applicants than they could handle. NHM already had the forms from the schools in Colorado, Washington and Minnesota so he began on those, while waiting to get the two additional ones. When all five were completed he asked his father for the money needed to submit them all. His father said, I thought only three were allowed. NHM told him that the college advisor had agreed to let him send in five because he didn't think there was much chance that he would be admitted to either Cornell or Rutgers. His father didn't respond, but he wrote the checks and NHM took the materials to school the next day. When they were all processed he mailed them off with his father's checks. Then he sat back as everyone did and waited to get accepted somewhere, anywhere, so long as he could go to college.

In NHM's group of close high school friends, the first to be accepted was a boy who wanted to become an engineer although his strengths were neither in math nor physics. He was admitted to a school in Pennsylvania well known for its engineering program and he was the envy of all. On top of that, his school sent him a small pocket-book with all the kinds of details an entering freshman would have to know. He carried it with him every day and let us read parts of it as time permitted. It was the coolest thing we had ever seen. Then NHM was admitted to the school in Colorado. Nothing like the pocket book came along with the letter of acceptance, although the mailing was precisely in the kind of large envelop rumor had it meant an acceptance. A small envelope was said to always be a rejection, a single sheet that said too bad. Within a few days the small envelope came from Rutgers. Both of the other two schools that NHM had chosen also accepted him. Nothing came from Cornell consequently he assumed they might not even bother to send rejection letters to applicants with his qualifications. NHM went into high gear then trying to decide which of the three schools he should attend. He found it difficult to rank them. He asked his friends what they thought. One knew that Colorado was a mountain state and thought it unlikely to have farms. Another said that Washington was famous for being on the Pacific Ocean. From what NHM read it seemed that Minnesota might be too cold for agriculture but he knew that there were farms in Canada so he thought it still might be a good possibility. Why didn't someone alert him to rank these three schools beforehand NHM asked himself? Now he was under the time pressure of having to let the schools know whether or not he was going to accept the place they offered to him.

NHM was grappling with his decision one Saturday morning, home alone at the time, when he heard the family dog attack the mail delivered through a slot in the vestibule of their house. It was a daily event and the closest person always rushed to retrieve the letters before they got all crumpled up by the dog's attack. NHM's father was very concerned always that his letters might be damaged so NHM rushed down to save the day. About a dozen letters had been delivered most of which had already been slightly chewed. He tried to flatten each of them out and to neaten up the pile. All of them looked like the kind of things that arrived for his father every day. He went into his father's office to put the pile on his desk. There he noticed by chance that one of them was addressed to him. It was from Cornell and was the typical small rejection type envelope. He took it up to his room to see what it said and to think of how he would tell it all to his father. What the letter said was a big surprise. It wasn't a flat out rejection. Instead it said that they could find no basis for his interest in agriculture in his application and asked first if he would outline for them "as specifically as you can your reasons for applying to the College…". The same sentence went on "..and also to indicate whether you would find it congenial and feasible to work on a farm this summer to get a start on the practical experience". That letter was dated April twenty sixth 1955. NHM took it to mean acceptance was likely and drafted his response as quickly and carefully as he could. On May twelfth, another letter arrived from Cornell. This one said, "I am glad to inform you that we are recommending your admission to the fall term provided you complete a full summer's experience on a good general farm". It went on to give instructions directing to whom NHM should write to get help in finding a farm where he could complete the practice requirement. Cornell University then sent him the official "ACCEPTANCE FOR ADMISSION" notice dated May sixteenth 1955. All that was required from their perspective was a forty five dollar deposit, a vaccination certificate, completion of a "Student Health Record" form, a final record from his high school, and, oh yes "contingent upon your completing the farm practice as specified by the Admissions Committee of the College of Agriculture."

With this in hand NHM told his parents that he had been accepted at Cornell University and explained to them what the "practice" requirement meant. He told them that students who did not come from farm backgrounds were required to work on farms each summer beginning before the freshman year, and to take a test on farm skills when they first arrived on campus in the fall. Each year thereafter they had to write a detailed report about their summer farm work and would be awarded points on the basis of their accomplishments. For satisfactory performance one point was awarded for each week of farm work. In order to graduate in Agriculture from Cornell you had to achieve a certain number of farm practice points in addition to meeting all other academic requirements. NHM explained to them that he had accepted this requirement and would be leaving as soon as possible after being graduated from high school. He explained to his mother that Cornell was an Ivy League school that was very hard to get into and told her he wasn't sure how he had been given a place there. She immediately called everyone she knew to tell them he had gotten into an Ivy League School, leaving out anything to do with agriculture. His father had nothing at all to say.

At school the next day NHM told his friends of his good luck. Word spread quickly. The college advisor called him in to get details. He said it was very important for the school to know how many students were admitted to Ivy League schools and told NHM that there were now seven going to Cornell. He gave NHM their names. NHM knew them all. Two of the others, Eddie Pilot and Dorothy Junglause and NHM had scored as the top three students in one of their math courses. Eddie's hobby was math. He was headed for the Engineering Physics program at Cornell where he quickly rose to the top of his class. Eddie lived near NHM and had been up to his room looking through his microscope. He wasn't surprised that NHM was going into life sciences at Cornell. NHM's friend Sig, the first one to get accepted said he always wanted to go to Cornell but didn't apply because his grades were not high enough to get into their engineering school. Eventually he also got to Cornell. Midway through his undergraduate years he discovered that he was really better suited for liberal arts than engineering and decided that he wanted to become a lawyer. Cornell accepted him into the law school and he loved it. Very few people knew that NHM's goal was to become a farmer. Going to the Ivy League was all they needed to know. The rest was of no interest to them but it was to NHM. NHM had to scramble to catch up on what agriculture was like at Cornell. He considered his chances of getting in so insignificant that he never had bothered to really examine the Cornell program. Once they sent him all the details about the colleges, majors, institutes, rankings, student body, living arrangements, location in the Finger Lakes region, and so on, he was able to get started. He read it all like a Bible. It made his head spin. He decided he would be an agronomy major although he had never heard the word before. It had something to do with soil physics. Maybe he could use physics in farming he thought. NHM knew that nobody would have any idea what agronomy might be so when asked what he was going to study at Cornell he always said agronomy. NHM cannot recall a single person asking what the hell that might be. Rather they all nodded politely as if that made good sense. NHM's mother said it sounded important. She was thrilled to have someone else's accomplishments to speak about beyond his father's. NHM's father was her only topic of discussion until then.

Cornell, it turned out, was a place of obstacles. How could anyone have known that without having to face them? The sweet success of becoming an Ivy League student quickly turned into hard work and sweat. Most of the students admitted (seventy percent) didn't make it to graduation. A few committed suicide along the way. NHM actually witnessed one jump to his death into a gorge one year when he was driving back to his apartment. That's putting the cart before the horse however. Let's start at the very beginning: the farm practice experience before NHM's freshman year. The Dean of the College of Agriculture put NHM in touch with Professor S. R. Shapley, the man who supervised the practice requirement for the college. He in turn put NHM in touch with Mr. John Boyd, who was in charge of farm placement in the New York City area working through the New York State Employment Service. Mr. Boyd placed NHM in a program called the New York State Farm Cadets. Finally, as a Farm Cadet NHM, would have his chance to work on a "good general farm…that would provide a better background for optimum understanding of the agricultural courses at Cornell, and count toward the farm practice requirement of the College." At seventeen this was the most exciting thing that had ever happened to NHM. He could already see the blue skies and the green fields. Getting out was almost within his grasp. The Farm Cadet program informed NHM that he was assigned to work on a small family dairy farm in the Adirondack Mountains not far from the Saint Lawrence River that separates N.Y. from Canada. NHM was sent a list of things that he would need to take with him, clothing, shoes and the like, and the name of a New York State County Agricultural Agent who would make all the local arrangements and introduce him to the farmer for whom he would work. The farmer would provide food and housing and a small stipend. NHM would work on projects of the farmer's choosing. It all made sense but it didn't play out exactly as anticipated.

NHM wrote to the county agent to get his assignment but the agent failed to respond. Finally he called the agent and learned from him that he had found a place for NHM but didn't have the details finished. The agent suggested that NHM let him know when he planned to arrive and said he would meet him and introduce him to his host. Together with his father NHM went to purchase the necessary work clothing and a suitcase large enough to contain it all. His father told him that he would have to travel on a weekend so that he could be dropped off at the bus terminal with his bag. NHM checked the bus schedules and found that he would have to leave on a Saturday afternoon in order to get to the small town where the agent was located by Sunday morning. NHM called the agent to see if it would be OK with him for NHM to arrive early on a Sunday morning. The agent said yes, he thought it would be, but he had to check to make sure that the farmer would also be free. He said that NHM should plan on it and that he would let him know only if they had to make another arrangement. So NHM's father bought the one-way ticket, gave him twenty dollars, and dropped him off at the Port Authority Bus Terminal in New York City on a warm humid spring day. NHM carried his bag to the bus and watched it placed in the cargo hold. He knew there would be many stops made during the trip and that he would have to change to another bus at one of them in the middle of the night. His bag would be automatically transferred so he had only to climb aboard the second bus and go back to sleep. NHM's destination was the last stop for the second bus so there was no chance that he might sleep through the place he had to get off. It all worked like a charm although it got much colder at night than NHM had anticipated and he was only wearing a short sleeve shirt when he departed on Saturday. There was no way he could get to his bag that had been checked through to the final destination. Nevertheless he managed to sleep well and felt ready to go upon arrival early Sunday morning.

When NHM got off the bus he went to pick up his bag at the side where all the cargo was unloaded. He couldn't find it. He asked the bus driver to check if it hadn't yet been unloaded. He did. Together they searched inside but it was empty. NHM showed him his checked-bag receipt. The driver took it to the cargo office and the man there said it probably didn't get transferred at the stop as planned but would most likely be on the next bus due to arrive at mid-day. NHM checked the time. It was just after 7:00 a.m. He realized that the county agent would be looking for him within the next half hour so he went out onto the street to look for the most likely place they would meet. The sun was up but the place was deserted. There was a main street that was paved and many side streets that were dirt roads. The town appeared to have not yet awakened. Seven thirty passed with no sign of his contact. At 8:00 o'clock NHM became concerned. A few blocks away he found a small diner that was open. There were a few people inside having breakfast. NHM went inside and asked the proprietor if he knew where he might be able to find the county agricultural agent. He explained to him that the agent was supposed to meet him at the bus station but hadn't shown up. No, he couldn't show up the proprietor said, Mister so-and-so died yesterday. At first NHM thought it was a joke. He did, he asked? The few people eating there turned around to listen to the conversation. One of them said, yes he died unexpectedly yesterday. Does he have an office where I might get some information NHM asked? Yes he does but it is not open today and probably won't be for the next week they told NHM. Thanks NHM said. He was supposed to introduce me to a farmer I am going to work for this summer NHM explained. Might it be any of you he asked. They all shook their heads no. NHM went back out onto the street to see if anyone else might be the candidate host. A pickup truck with an elderly couple drove by. Could those be the people he thought, but they kept on going right past him. About ten minutes later they circled back so NHM flagged them down. I am looking for a farmer I am supposed to work for this summer, NHM said, could it be you? Yes I think so; we are looking for the county agent the farmer responded. I was looking for him too, NHM said. The people in the dinner told me he died yesterday. We did not hear that he said. No wonder we couldn't find him his wife added. NHM told them his name and asked theirs. I am Leslie Fleming the farmer said, get in and we will go to the farm. Where's your suitcase he asked. It is lost but the Bus Company said it would probably arrive on the mid-day bus NHM explained. Do you have anything else he asked? No I don't NHM said and they drove off to the farm.

Shortly after they got underway it began to rain. It rained on and off for five or ten minutes. There appeared to be no other cars traveling on the same road as they wound their way besides fields and forested places. Then they came upon an accident on the other side of the road. A convertible with its top down was lying on the passenger's side just off the road. A small group of people were milling around. The farmer stopped his truck and we got out to see what had happened. A young woman was sitting with a man nearby crying. She kept repeating Patty, Patty, oh my God if anything happens to Patty my mother will never forgive me. NHM walked over towards the car that had rolled onto its side. There lying beyond it was a young girl. Part of her scalp and brain was nearby. She had been killed. Just then a Priest arrived and went to administer last rites to her. It was horrible. NHM couldn't believe the tragedy of this place. First the agent died before NHM arrived, then this young girl was killed in a freak accident. NHM's farming career wasn't off to a great start. The final blow, however, came on the farm. To get there they drove from the first small town where the farmer had picked NHM up to another even smaller place, continued eighteen miles beyond that then turned off onto a dirt two tire track road and climbed into the mountains to an isolated clearing. There stood a small farmhouse, an old barn, with adjoining barnyard, a small chicken coop, a small field, and some forests. The farmer said they would have to cut some hay for the rest of the day until it was time to milk the cows. He asked if those were the only shoes NHM had. Everything else is in my suitcase NHM told him. I can lend you a pair he said until yours get here. Thanks, NHM said. Try these on he suggested as he offered me a pair of what looked more like dress shoes than work shoes. They were too small but NHM could just squeeze into them by opening the laces as widely as possible. They'll work, he said. Good the farmer said, they used to belong to my son-in-law. We have some other things of his here that you might be able to use as well. Last year he died in the bathtub here, he said in a rather casual way. The radio fell into the tub and he was electrocuted. The farmer's wife became very angry that he would speak of it in that way. It was difficult for NHM to use the bathtub for the remainder of his stay on this marginal farm. He routinely went to bathe in a nearby stream instead. Twice on his journeys there he encountered what must have been a large bear although he never got a good enough view of it through the forest. As soon as he had saved fifteen dollars from his stipend, NHM purchased a single shot Winchester 12-gauge shotgun. It was with him whenever he had to go anywhere in the mountains alone.

Fortunately before NHM had gotten to the farm he had learned some things about guns. At thirteen his father had given him a rifle for his Bar Mitzvah rather than the traditional fountain pen. Many in the family never forgave him for doing so, but NHM took advantage of the gift and learned how to use it. Then his father repeated his generosity shortly after NHM's sixteenth birthday. NHM needed an operation on his neck to remove a possible tumor that turned out to be benign. The anesthesia nearly killed him. It took a week for him to regain consciousness. When he got home to finish his recovery it was almost Christmas, a holiday they celebrated as a gift-giving occasion rather than a religious one. NHM was hoping for a shotgun but saw from the size of the packages under the tree that Santa didn't bring one. But in fact he did. It came disassembled so the box was short! What a great surprise it was. When NHM fired it the next summer he realized it was an unpleasant weapon. It nearly deafened you and produced a nice black and blue mark on your shoulder. NHM more or less gave it back to his father before he graduated from high school, and never thought about taking it along to the farm the following year. What he bought on the farm was a much simpler weapon but just as lethal. The only three times he had to use it were right near the farmhouse, not in the forest where true dangers lurked. In the first case an aggressive rooster had killed one of the few hens that provided the family with eggs and the farmer decided that we would have to kill it. The job was assigned to NHM. Shoot off its head he said. NHM chased it into the coop and carried out the order. The farmer retrieved the bird, gave it to his wife and she served it for diner. The second time was later in the summer. A small field of oats that the farmer grew to provide food for two draft horses he owned was invaded by a large flock of blackbirds. Quick he said, get your gun and shoot into the flock. NHM's first blast caused the flock to rise off the grain. The second knocked down a small number of birds out of the huge population. The farmer captured a wounded one and tied it by a leg to a fencepost at the edge of the field to scare off others. A third blast drove the flock from his grain field over the trees into the forest. The final time NHM used his shotgun was when they were working to cut the brush that had grown over the dirt road leading into or out of the farm. It had become almost impassable. He and the farmer worked in parallel along the two sides of the road from the farmhouse toward the paved road two miles away. At one point NHM stepped forward into a hidden beehive disrupting it enough to cause part of the swarm to rush up his pant leg and deliver a half dozen stings. They both retreated to the truck for protection and drove it as close to the farmhouse as possible. They ran in and the farmer got some newspaper and matches and he told NHM to get his shotgun. They then drove back to the location of the hive. The farmer set a small fire around it. He then asked NHM to shoot into the hive to incite the bees to fly out and thus be burned. It seemed to work. NHM was sure this is not what Cornell had in mind for farm practice. Could this be it, he thought? Impossible. None of it was included in NHM's farm practice report.

NHM's suitcase never did arrive during the summer. After two weeks of making do with the dead man's clothing NHM was forced to call his father and ask him to send money for replacement goods. His father asked him if had a way to get to a store nearby where he could purchase what he needed. NHM told him that the closest place was twenty miles away and that he might be able to go with the farmer there on one of his infrequent shopping trips. What will you do in the interim NHM's father asked? I'll make do NHM said but it's not getting any easier. The farmer's wife washes my clothes once a week NHM told him. Maybe I should drive up there his father said. He then went on to tell NHM that the family was going to leave New York. Their house was going to be sold, and his Manhattan Office closed. He had accepted the position of Director of Instruction at a mental hospital in southern New Jersey. He said he wouldn't mind taking a break from all the details of the move to come up and get NHM some new clothes. NHM thought it was a bad idea for many reasons but his father insisted. Three days later there came a call. They were in the neighboring town and needed driving instructions to the farm. NHM gave them the details and said he would ask the farmer when he might be able to take a break from work to go and get the new clothes. The farmer suggested the following afternoon so NHM made the arrangements with his father to pick him up then. His father and mother arrived on time. After brief introductions all around, the farmer's wife told then that NHM did not eat much. Then the farmer himself mentioned that NHM had probably told him all about the dead county agent and the girl killed in the accident the day he picked NHM up. On the way to purchase the replacement clothes NHM's father said, "you didn't tell us about any of that". I guess I was too occupied with other things at the time, NHM replied. Nothing more was said. NHM wondered why the farmer mentioned that to him. Perhaps it was all he could think of to tell a doctor? As soon as they arrived in the small town they quickly purchased the replacement clothing. NHM gave his father the claim check for the lost bag so he could file a claim to recoup the cost of getting the new stuff. Before they returned to the farm NHM's mother suggested they stop for a milkshake or ice cream. How unusual NHM thought. She had never done that before in his memory and his father was not one to stop for anything other than gas on trips. They walked to a drug store that had a fountain where you could get these things. NHM realized his mother must have thought that there wasn't adequate food for him to eat on the farm. Indeed there wasn't much but he could manage all right with whatever they had to offer. At the time NHM assumed that the Farm Cadet program must have provided the Flemings with some funds to cover his expenses, but it probably hadn't really worked that way. What it must have been was a program meant to find laborers in New York City who could be exploited to do farm work and at the same time get them out of the city. Cornell bought into this program probably as a means of discouraging students who really weren't headed into farming from entering the College of Agriculture rather than one of the other Cornell Colleges. As an endowed college, Agriculture cost much less to attend than any of the others did.

NHM could make do but had to admit that the rooster they ate that day tasted precisely like rubber bands! So NHM drank a milkshake and the three of them then motored back to the farm. After unloading the items to his room upstairs his mother said it was getting dark and suggested that they depart before too long so that it wouldn't be too dark when they had to drive out along the farm road. After saying goodbyes they got into their car. Uh-oh, it did not start. Not a sound of ignition or anything else could be heard. Repeated attempts to start also failed. Perhaps the battery became disconnected the farmer suggested. Nope it wasn't that. Under the hood nothing could be seen to explain the problem. By then it was dark and the mosquitoes had started biting. NHM's mother got panicky. The farmer's wife suggested they could all stay overnight upstairs in NHM's room. She had some blankets they could put on the floor. No thank you they said without a second's thought. Could we get a ride back to town they asked? Well not tonight the farmer told them. There were things he had to do in preparation for work in the morning. Could they call a cab from town NHM asked? Maybe one would be willing to drive all the way. You'll have to pay for both directions the farmer said. Fortunately a phone connection was available and arrangements were made. NHM's parents left the car and returned to their room in town. The next morning NHM's father drove out with a mechanic. NHM was working in the barn and therefore couldn't see how the problem was solved. When he heard the car start NHM went out of the barn. His father was behind the wheel. He waved from the window and drove off. Next NHM saw them was at their new home in southern New Jersey after he returned from the farm at the end of the summer.

What NHM did on the farm is a book in itself. It was basically nothing more than slave labor placed under the supervision of an impoverished farmer who had virtually nothing that he could teach about modern farming or farm life. The entire farm community in the area of Mr. Fleming's farm lived near the starvation level. No single farm had the equipment it needed to do all the things that had to be done. Instead they used a kind of barter system that involved both machines and labor. In exchange for the use of a chainsaw to cut firewood in preparation for winter, NHM was sent to work for the man who owned it. Three days of NHM's labor was the cost for an afternoon's use of the chainsaw. NHM worked for a week on a threshing machine in exchange for having the small amount of oats that were grown on Leslie Fleming's farm cut and threshed. The Fleming farm had but three pieces of equipment: a pickup truck, a tractor, and a milking machine. All three were obtained, the farmer told him, after the Second World War as part of a government program to keep family farms working. NHM had no idea how he acquired the twenty five cows that he owned. No records were kept of anything. NHM tried each day to put an entry into his farm diary from which he eventually produced the required report for Cornell. The small amount of milk produced by Mr. Flemming's cows was picked up every other day from the roadside and disappeared to a cooperative somewhere that NHM never visited. The only other source of farm income was a few calves born during the summer. The males were sold to a dealer for a price based upon weight. Only two females were born while NHM was working there. One was kept, the other sold to another farmer. The same work routine was followed every day. The farmer's wife tended a small vegetable plot. Towards the end of summer part of it had to be plowed. To do so they used one of the two draft horses. NHM rode upon it taking directions from the farmer who walked behind with a primitive wooden plough. The year was 1955 but the image was that of farming in the 1800's or earlier.

If something became broken that could not be fixed on the farm the item would be taken to someone else who was able to make the repair. Usually the barter system was used to pay for the work with one exception that NHM recalls. The tip of a hay-cutting tool struck a rock and cracked in half. It had to be taken to a welder. The man who repaired it asked fifty cents for his work. The farmer had but thirty cents with him so the welder accepted that. In the group of farmers that NHM came to know there was one who appeared to be much better off financially than the rest. The first time NHM met him both he and Mr. Fleming went to work at his place for a day to pay a debt that had been incurred before NHM had come to work on the farm. NHM was shown around the place before being assigned the tasks that he would have to do. In one of the barns there was an enormous draft horse, much larger than any that NHM had ever seen. After lunch NHM asked Mr. Fleming if he had seen the beast. Yes he said as they walked over to take a look at it before starting work again. They both marveled at the size of the animal's head. I sure would hate to have to feed that, Fleming said to NHM. Owning it was clearly a luxury. Even more impressive however was a very large hay storage barn. A group of people was assigned to remove all the hay from it during the afternoon. The bales were piled in a field nearby. Then they swept the entire place clean. The upper floor of the barn had a beautiful wood floor finished as if it were in a house. When the entire place was empty and clean a tank truck drove into the barn and whitewashed the walls and ceiling carefully covering the upper floor before the spraying began. What's going on NHM asked the crew boss to whom he was assigned? It's for the barn dance this weekend he said. Aren't you coming? Don't know yet NHM said. That evening NHM asked Mr. Fleming if he could drop him off there on Saturday night for the dance. Fleming agreed but told NHM he would have to get back by himself and be ready for chores on Sunday morning. That was one of the two days that NHM went off the farm during the summer, not counting his trip to buy replacement clothes.

When NHM arrived at the barn dance the place was jammed. They had a country western kind of band playing. It was clearly a major social event of the summer. There was a fifty cent admission charge and you could buy sodas if you wanted to inside. The girls wore dresses that could have been in old classic movies. All the farmhands looked like western dudes from Texas. For a student of sociology or culture the farm experience couldn't have been beaten as a place to learn about rural America. For a person who wanted to learn how to be a farmer it was a lesson in what failure would bring. No one that NHM could identify there had ever been educated in anything. They seemed to be living on information handed down from generation to generation that perpetuated the primitive conditions of their ancestors. None of this went into NHM's farm journal. Instead he focused on exactly what the work consisted of, and incidental things he thought to be of a 'scientific" nature and appropriate. For example, he included details about a thin tiny animal NHM found one day when cleaning out manure and urine from the barn. It was about the size of a hair thrashing about in the fluid. What's this he asked Mr. Fleming? The farmer looked down and said it was a hair snake. It wasn't any kind of snake at all. Rather it resembled a larger version of something NHM had found in his fish tank, some kind of invertebrate. Where does it come from NHM asked him? From nowhere he said. What do you mean nowhere? It just comes from nothing, Fleming said. NHM couldn't believe his ears. No living thing comes from nothing NHM said to Mr. Fleming. It must come from some tiny eggs too small to see without a microscope NHM said. Then they went back to do their work. That evening at the dinner table the farmer recounted what had happened to his wife as if it were something new to think about. Neither of them asked NHM anything more about it, nor did they discuss it. Could they have believed something living comes from nothing, a kind of spontaneous generation NHM wondered? This is the kind of thing NHM wrote into his farm journal. In the end, for nine weeks work NHM got six farm practice points and a very low grade on the practical test that they gave him at Cornell. NHM didn't know the difference between alfalfa and hay grass such as Johnson grass. The hay he had worked with was some kind of grass but God knows what kind. NHM doubts any of the farmers he met knew either. NHM couldn't back up a wagon attached to a tractor into a marked off area. They had no wagon on the Fleming farm where he worked. NHM had the feeling that the chips were stacked against him in this agriculture program. Was it possible that someone couldn't do agriculture coming from the street in New York City he wondered? But NHM would not be defeated that easily. Not when his whole life-plan rested on farming.

At the end of the summer NHM packed up his few belongings including his Winchester Shotgun, and had a discussion with the farmer before leaving. Mr. Fleming told him that he would not take another New York State Farm Cadet because, "they don't know nothing about farming and can't do as much work as a local farm boy can". NHM purchased his bus ticket and headed back to his parents' home, now at a state mental hospital in southern New Jersey, known as Ancora State Hospital. By the time he got there it was less than a week before he had to depart for Cornell. NHM went over the list of essential things that Cornell recommended he bring with him and packed those he already owned into a old army footlocker that his father had kept from World War Two. This he sent to Ithaca addressed to the men's dorm where he would live for the first year. The things he did not have he assumed he would purchase when he got to Ithaca. NHM's mother insisted however that he buy them before departing and take them with him. She urged NHM's father to take NHM to Philadelphia, the closest city to their hospital home, and purchase the needed things. His father agreed so they drove into the city one afternoon but their destination was uncertain for neither of them knew where the shopping districts were or which stores would be appropriate. Eventually they came upon a place where some men's shops were located, parked and searched the storefront windows until they found a store that appeared to carry appropriate merchandise. Inside NHM read off his list of items to a salesman who pointed out what he believed would be suitable. There was of course no way of knowing whether any of it was appropriate but as a matter of expediency NHM settled upon whatever the salesman showed them. After they returned home to the mental hospital NHM packed it all into his suitcase. He never gave it another thought until he became settled in at Cornell. There he saw what the appropriate clothing should have been. What he had purchased in Philadelphia might have been ideal for a middle-aged man living in a city who hadn't quite made it yet to suburbia, not an Ivy League college student living in one of the coldest snowy places of the northeast. Fortunately NHM had a black suit from his days in the high school orchestra that he could wear to important events but even that wasn't as versatile as he had hoped. When NHM established a small band to play at fraternity parties he had to rent a tuxedo. The money he earned from playing became earmarked for purchase of necessary clothing and other living expenses. The first thing he bought was a surplus army parka with a fur hood. It was standard gear for blizzards all winter. Then NHM bought one new suit every year with matching shirt and tie. By the time he graduated he could pass for any other student coming from the kinds of wealthy families that sent their children to the Ivy League after prep school.

The trip from the mental hospital in southern New Jersey to Cornell University in Ithaca was more than a journey of several hundred miles; it was more like leaving one planet for another. NHM's father dropped him off in the late afternoon at the train station in Philadelphia for the overnight journey to Ithaca. He boarded with his suitcase at about dinnertime. The train arrived in Ithaca the next day at about 7:00 a.m. It was a clear brisk day in September. When NHM stepped off the train the setting almost knocked him off his feet. The most beautiful place of his dreams had come true. The trolley had landed! Trees, mountains, a blue lake were all completely unexpected. Other students who got off with him knew the drill well. Friends either met them or they boarded taxis in groups and headed up the hill to the campus. The cabs made the rounds of dormitories, fraternities and sororities dropping off students at each stop. NHM got off at the men's dorms and made his way to the room where he would spend his first year. Most of the other residents had not yet arrived. He found his footlocker in a storage room, dragged it to his room and got everything unpacked before his roommate appeared. Then he went out to take a look at the campus. The views were spectacular. The buildings looked like palaces. Everything he saw was like looking through rose-colored glasses. Even the people seemed friendly! Then and there he decided, he was going to stay there forever. I am never going back, NHM thought. They will not defeat me here even if I don't know anything. This is way too important to me. Here is where I am going to find my place in the world and establish my life. Forget the past.

During the first week of orientation at Cornell, NHM learned what students like his old high school friend are taught at prep school in addition to academics: they learn what it's all about. When they got to college, there were no surprises. On the way to hear the Cornell president's welcome to the freshman class, a former prep school student who lived in NHM's dorm walked over with him. On the way he said, I'll bet he's wearing a red tie. Perhaps so, NHM answered. Wait until he gets to the, "look to the left of you, look to the right of you, part about how only three of every ten will graduate, he said. Have you heard his talk before, NHM asked him, is it always the same talk? When I decided to go to Cornell my prep school advisor arranged a meeting with three Cornell students who had attended our school. They clued me in on everything. How useful, NHM said. Yeah, he replied, didn't they do the same at your school? No they didn't, NHM told him. How many came to Cornell from your school this year, he asked? Seven NHM responded. Wow, seven and they didn't arrange anything. You didn't get your money's worth, he said. I guess not, was all NHM could say. When he got back to his dorm room he asked his roommate if he had gotten all this stuff about Cornell from his prep school. What stuff, he asked? Well the things that you get told during orientation week, NHM said. Sure we did, didn't you, he asked? No I missed it, NHM said. Why did you come to orientation week then, NHM asked? Well he said, my father is out of the country on business, and my mother doesn't live with us anymore, so I came directly from school. They required that graduating seniors leave as soon as the college dorms open, so here I am. You might have noticed that I haven't gone to any of the orientation events, he said. No NHM hadn't noticed, he was too busy getting all things done before classes started. He had to take the farm practice test. He was required to take a math placement test because in his senior year in high school he didn't take the last math sequence in order to leave time for orchestra. NHM had to meet his academic advisor, as well as an assistant dean in the College of Agriculture who spoke to each student who came from New York City. This dean's job was to determine whether these students were really interested in Agriculture as opposed to the financial benefits of going to the endowed college. Prep school students were unlikely to be going to the College of Agriculture so NHM assumed they wouldn't have quite so many things to do during orientation week

Everything went smoothly with the exception of NHM's meeting with the Agriculture vice dean, Dr. J. P. Hertel. The moment he met Hertel he sensed that Hertel disliked him and distrusted him. Hertel spoke with a heavy German accent and walked with a cane, which made NHM very suspicious of him as well. What are you doing in Agriculture, he asked? Trying to learn to become a farmer NHM told him. "You did very poorly on the farm practice test" he said; "perhaps you have no talent for agriculture." NHM did not respond so he went on, "Are you a New York State resident ?" Yes I am NHM responded. What is this address then that you have given as a permanent address in New Jersey he asked? Well my parents have just moved there, NHM said. When, he asked? This past summer when I was working on the farm, NHM said. You mean before you registered here he asked? Yes, NHM replied. Then you are not a New York state resident, he said. I have lived in New York State my entire life, NHM told him, and in addition my parents own a property in upper New York State. That doesn't matter he said. Where you live at the time you register is all that matters, you could move into New York the next day, it doesn't matter, you will be a non-resident for your entire 4 years if you last that long he pronounced. This means you will have to pay out of state tuition he informed NHM. I will, NHM asked? You will pay tuition, he repeated, and if your farm practice points don't improve you will not be able to stay in the College. They will improve, thank you NHM said and he left. NHM's father became very angry when he related it all to him that evening on the phone. How much is the tuition he asked? I will find out and let you know tomorrow NHM told him. When his father heard the figure he seemed relieved. That won't be a problem. I am glad to hear that NHM said. I might be able to earn enough playing in a band he added. The next time I come home I will bring my bass back with me, NHM said. OK, he replied, but his father never did ask for money to cover tuition. Years later NHM learned that his father paid the tuition with money he was paid for a medical disability he had gotten in the army as a result of a shoulder injury he suffered while on a training mission. Tuition amounted to several hundred dollars a year in the 1950's. NHM definitely could have made that much playing in the band.

NHM's advisor in the Agronomy Department, a man named Dr. Sterling B. Weed, also asked how he came to choose Agriculture, and Agronomy in particular, so NHM explained it all to him. Weed seemed to accept his explanation but then pointed out to him that agronomy courses required so many precursors be taken first that NHM would not be able to take any agronony classes for at least two years. Together they planned the first year's courses and outlined everything else that NHM would have to take in order to satisfy the requirements of the major. NHM registered for everything then purchased the books and materials needed for his courses including a small nylon windbreaker jacket that was recommended for fieldwork. It was an ideal weight for the season so he began wearing it for everything in place of the coat he had bought in Philadelphia. By week's end he switched back however. The nylon jacket was stolen from a coat hook where he left it while eating dinner in the Student Union. Not everything was perfect at Cornell. The following week was the start of class. On the very first day he went to the math class that he had been assigned to take in view of his entry deficiency. When everyone had settled in the instructor said is Mr. Mendelson here? Yes I am NHM replied. Would you come up please he said. What could it be he wondered? You scored very highly on the placement exam, so you have been given credit for this course without having to take it, he said. Here is the sheet you should take to your Advisor. You will be able to select another course instead of math. Thank you very much NHM said and he left. This might be easier than I thought NHM said to himself on the way over to his Advisor's office. When Weed saw that NHM had tested out of the math course he became very enthused. Let us see what else fits your course schedule that you might like to take. Surprisingly there weren't very many options at the time NHM had free. He chose a course in meteorology assuming it might teach something about the microenvironment around plants. Unfortunately it didn't. The focus was on atmospheric physics. The Professor told him he had never seen a freshman attempt it before, but would let him stay in if he wished. NHM decided to give it a try. It was very difficult but he survived, and he learned some things about how complex the dynamics of the atmosphere are. He learned that Ithaca had the lowest light intensity east of the Mississippi, and why on what he would have said was a totally clear sky over the campus, the cloud cover would in the expert's opinion be ten tenths covered. It appeared that NHM had never seen the true blue color of a sky that was totally clear. When he thought of it, NHM realized that most of the time he never took his eyes off the dangers of the street. Looking at the sky had not been part of his life-style.

It didn't take long for NHM to become settled into life in the men's dorm and to adjust to the routine of college life. One thing that NHM was totally unprepared for were the kinds of devious things these very smart students had learned to do to one another in their prep schools. They of course recognized the opportunity he provided them, and it wasn't long before they duped him into doing a little nothing that turned out to really be a big something. Just after dinner one evening NHM was working in his room on some class material when a student who lived across the hall on the same floor came in. He said, Neil you can play the trumpet, can't you. Yes a little, NHM replied. Come over here to my room he said I want to show you something. The window in his room looked out onto a courtyard that was surrounded by six men's dormitory buildings. The lights appeared to be on in virtually every room and the occupants were all looking out as if something happened in the courtyard. NHM couldn't see anything they were looking at. He asked, what's going on. Everybody's waiting he said. For what, NHM asked? Your signal he said. What signal? Look, just blow charge on the trumpet he said and he handed NHM an instrument that belonged to someone on the floor who couldn't really play it. No, I'm not going to do that NHM told him, there are people studying. Look, he said, no one's studying now. Just blow it, go ahead, no one will be angry. So NHM took the horn and he blew it. It surprised him how good it sounded. Then within a moment he noticed that there wasn't anyone looking out of the windows anymore. A few moments later they all came charging out into the courtyard and began running onto the campus. Where are they going, NHM asked? It's the first panty raid of the year he was told let's go. NHM quickly gave back the trumpet and went into his room to ponder what he had triggered. From his window he could see the crowd running toward the girls dorms. Oh shit, everyone's gone he said. It occurred to him that the cops might soon arrive at the dorm to find out who started the whole thing, and NHM was probably the only one left in his dorm building. He decided he better get out too. So, he hurried to catch up with the rear of the crowd. Before he had reached them however, a police car came careening around the corner and nearly ran him over. The cop stopped and asked him where he was going. NHM said to see where the crowd is headed. Look, the cop said, just go back to your dorm before someone gets hurt. As soon as he sped off NHM continued to the girls dorms. The men were milling around the buildings shouting to the girls who were all at the windows waving and shouting back. The doors had been locked and barricaded so no one could enter or leave. Occasionally a window opened and panties were thrown down, which resulted in a mad scramble below. NHM thought, wow what a trophy that would be to mark the raid he triggered. Try as he might he couldn't get close enough to get one. The cops arrived and tried to break it up but the girls shouted at them to leave them alone! When the police nabbed a couple of the most aggressive kids, the raid petered out, the crowd dissipated, and NHM went back to his room. The student newspaper carried an article about the raid the next day and simply said someone blew charge on a trumpet, and that was the predetermined signal for the start of the raid. Where the hell was NHM when that was decided? How could he have been so stupid? Fortunately there were no serious consequences. NHM certainly didn't want to get thrown out of college on a trivial matter and ruin his chance to learn how to become a farmer.

Towards the end of his second year NHM got a note from the infamous Dr. Hertel asking him to make an appointment to see him. He knew the news would not be good. Hertel informed him that he had fallen so far behind in farm practice points that no one had ever been able to make up the deficit in time to graduate. He said he knew NHM's course grades were good and that he had taken some very hard courses but suggested that perhaps agriculture wasn't for him. Have you considered going elsewhere or switching colleges he asked? No I haven't NHM told him. This doesn't make sense NHM said. Several professors in the college from whom I have taken courses have suggested that I think about going on to graduate school, NHM told him. I am not saying you're not qualified to do that he said just that the Agriculture College has certain other requirements that you must meet. I will definitely meet them NHM told him and left. He went directly from Hertel's office to see one of the professsors in the Botany Department from whom he had taken a great plant physiology course. The professor knew NHM was very interested in genetics and in plants and he had been very encouraging about NHM's continuing on to graduate school after finishing a bachelor's degree. NHM told him the problem and he immediately suggested a solution. Switch to the Botany Department he said and you will be able to substitute research practice for farm practice. The Botany Department he explained is part of two colleges, Agriculture and Liberal Arts. Botany majors can be in either but those in Agriculture have to petition to be admitted to the Department in order to assure the university administration that it wasn't just a move meant to save tuition money. He asked NHM to simply write a letter to him asking to become a Botany major, and explaining that his goals were to continue work in graduate school focusing on genetics and perhaps plant genetics as well. He assured NHM the Botany faculty would grant his request and once done, the College would have no choice but to allow him to substitute research practice in lieu of farm practice. NHM wrote the letter he asked for and delivered it to him the next morning. When he presented it the professor said do not worry, I will take it to the Botany Majors Committee in the Department and write to you as soon as it has been voted upon. Everything he said was true. Here is the letter NHM received from him:

NEW YORK STATE COLLEGE OF AGRICULTURE
A UNIT OF THE STATE UNIVERSITY OF NEW YORK
CORNELL UNIVERSITY
ITHACA, NEW YORK

DEPARTMENT OF BOTANY
PLANT SCIENCE BUILDING

May 7, 1957

Mr. Neil H. Mendelson
310 Triphammer Road
Ithaca, New York

Dear Mr. Mendelson:

I am pleased to inform you that the Botany Major's Committee has voted to accept you as a Botany Major with substitution of Botany Practice for Farm Practice permitted. Among the responsibilities placed upon you as a result of this action are the completion before graduation of the remaining required Botany courses (Botany 117 and 124) and Plant Breeding 101, the fulfillment of 25 units of Botany Practice in approved situations, and the attainment and maintenance of a high scholastic standing.

I would be glad to discuss these and other matters with you at any time.

Yours sincerely,

Donald G. Clark

D. G. Clark

DGC:sp
CC: Professor J.P. Hertel
Professor S. R. Shapley

NHM went to thank him after he received it. Dr. Clark told him he would arrange for NHM to work over the summers on plant genetic projects conducted in the Department of Plant Breeding. This man, Professor D. G. Clark saved NHM's career at Cornell. He became NHM's advisor and made good on all of his promises. Instead of going off to a farm each summer NHM stayed at Cornell and worked in the experimental plant genetics program on campus and at nearby agricultural experiment stations. The chairman of the Plant Breeding Department, Dr. R. P. Murphy, arranged for NHM to get the necessary funds, and assigned him each year to the particular projects that he would pursue. The first year NHM worked on a grain-breeding project with one of the top wheat geneticists in the country, Dr. Neal F. Jensen. The second year he worked on several vegetable-breeding projects under Dr. Henry Munger, the very same man responsible for developing the tomato variety that cannot be destroyed by shipping. Unfortunately humans who do not eat rocks cannot eat it, but it must have been great for the tomato industry. Before he knew it, NHM had taken every course that was available in genetics and became convinced that he ought to go on to get an advanced degree, and possibly even become a professor! He convinced himself that genetics could be a vital part of farming. NHM hoped that maybe both would work together to assure that he would never have to go back to life in the city. No, the likes of Hertel did not defeat NHM, but without Dr. Clark NHM might not have been able to hurdle past him on the way to graduation from Cornell and the rest of his life. Professor Daniel Clark enabled NHM to pass the point of no return on his escape journey although he knew nothing of the details that made it a necessary journey.

Life at Cornell was full of surprises. Although the students were very smart, few, if any, were intellectual in any real sense. Most of the men were heavy drinkers. Cheating was rampant. Social life focused on getting coeds into bed rather than meeting someone who might be an appropriate spouse. It was difficult to know whom to trust. Many of the professors were inaccessible. Teaching a course was a terrible burden for them, a distraction from their primary objectives. Grades awarded were uniformly low. Not many of the undergraduate students were happy there. Official policy was to let those who could keep their heads above water do so while the rest were allowed to sink as soon as possible. It was understood that there were at least ten fully qualified candidates for each student admitted, so if you failed you could easily be replaced. The rules were ruthlessly applied by the administration. A friend whose father died unexpectedly on the first day of finals missed all his remaining exams. Didn't matter why, he was failed in all his subjects and suspended. The terrible pressure took its toll on many students. Each year some students committed suicide. Others left terribly discouraged and for some it was really impossible to recover. At the other extreme there was a small group who appeared to be unfazed by all of the tortures. These survivors somehow got through it unscathed. NHM was one of them. It wasn't that the successful few avoided the onslaught. No, they took their share of the beating but it just didn't destroy them. NHM for one, found much of it amusing. That is hard to understand. NHM certainly didn't have the financial security most of the other students did, so he just couldn't walk out and become a vice president in the family's business. No, it wasn't that. It was more that, somehow, NHM just expected that he would be put through it and was surprised that it wasn't worse. Is this the best they can do, he often thought, how pathetic. The more he succeeded the bolder he became. He beat Hertel at his own game. He developed a group of close friends. He got to know and trust a group of professors. His grades were reasonable; his band was functioning well, kept busy working by the relentless sequence of parties that constituted the social scene throughout the academic year. It was a time to hunt for a wife and to have some fun as well. As a diversion NHM began to play tricks on others, to prey upon them in a way as others had preyed upon him, but things didn't always go his way in this, as you might expect.

Humiliation is the word for what resulted from one of NHM's first unsuccessful pranks. While living in the new fraternity house, a building that was constructed between his freshman and sophomore years, a group of pledges arrived on a trip from another university. They had with them a live piglet that was dressed as a baby and they were required to get it back with them alive and with pictures of it being held by sorority girls at various colleges. Let me have it NHM told them, I'll make the arrangements for you. When other brothers informed them that NHM was in agriculture they immediately relinquished the animal (it was of course the first time he had ever touched a pig). NHM carried it in its blanket to the sorority house next door, rang the bell and explained to the girl who answered the door that he needed help with a baby who had been brought by pledges to the fraternity house. He did his best to try to invoke a mothering instinct and it seemed to be working. Can we see the baby she asked? Certainly, it's right here, NHM said. By then two or three other girls had gathered around to get a look. NHM carefully removed the blanket from the piglet's head. No sooner did it see light than it jumped out of the blanket to the floor and scrambled away through an open door into another room. The sorority girls did not think it funny. Can I go and fetch it NHM asked, as defeated as one could be? Indeed, please do was their response. So he rushed into the next room. What a stupid thing to do. It was of course the living quarters of the housemother, an elderly biddy having a tea party at that very moment with several friends of like demeanor. Certainly they knew the animal was scurrying around the room but paid absolutely no attention to it. May I come in and remove the pig NHM asked? I'm terribly sorry, he said. The housemother responded with: "yes, remove it". That was not easy to do but NHM would have killed himself if he had to in order to catch the damn thing and get out of there. Fortunately he was able to grab it without ruining anything in the apartment. When he got back to the fraternity house NHM gave the pig back to the visiting pledges and told them the girls wouldn't touch it. Why not they asked? He certainly wasn't going to let them know that he had made a complete fool of himself so he simply said they claimed it smelled bad. That was that.

At the end of his second year NHM made some major changes in his life at Cornell. He reversed an earlier decision about not continuing after the first two required years in the Reserve Officers Training Corps. The Vietnam War was just starting but the language coming from Washington suggested that a major confrontation was in the making. There would certainly be a draft and NHM as well as everyone else his age had already been registered at their local Draft Boards. NHM reasoned that as a student of agriculture there was little he could claim that might help keep him in school rather than in the military and thus he thought it best if he was going to have to serve to do so as an officer. He went to speak to the people in the military science department to get the details and necessary forms to continue in the program. After digesting all the literature he made application requesting that he receive a commission as a Reserve Officer in the Army and be assigned to the Chemical Corps rather than a combat arms branch such as the infantry. He also indicated that he planned to go to graduate school after completion of his Bachelors degree. NHM's goal was to obtain a Ph.D. degree. After submitting the application he was informed that he would be accepted into the program contingent upon the results of a physical examination, and satisfactory completion of his current courses in military science. He was told that a four-year delay in the call to active duty would be granted to all new officers who were enrolled in a Ph.D. program. Upon completion of graduate work he would be given the choice of serving for six months on active duty followed by seven years in the active Reserves or two years of active duty followed by several years in the inactive Reserves. NHM's fraternity brothers thought he was insane to consider continuing in the military program. They seemed quite certain that they would never serve, and they were correct. How they avoided it he does not know. Nevertheless, NHM was somewhat nervous about the decision that he had made. He tried to focus on the positive things about it. His father had served in WWII. All Jews should know how to fight. Better an officer than a private. You get paid to take the advanced training. It might not be worse than the streets of New York.

NHM went to take the physical. It was a typical military physical. Each part was done at a different station. He reached the station where a physician listens to your chest to check your heart and lungs. The doctor seemed to be taking an awfully long time. He called another physician to listen as well. Then he asked NHM, "have you ever had any heart disease?" Not until this moment NHM responded, as confident as could be. NHM explained that his father was a physician and that he was sure that if anything had been wrong with him his father would have known it and so would he. I am going to send you for an electrocardiogram the examining doctor said. There is no cost. Is there a phone where they can call you to arrange an appointment, he asked?

Yes of course NHM said and gave him his number. The doctor said if you don't hear from them within a week call this number and tell them who you are. NHM wrote it all down and walked out of the exam as if he would drop dead the next moment. He moved as if he were walking in molasses not air. When he stepped off the curb he didn't let his foot drop into the street as in normal walking. Rather he touched down as if stepping on eggs while trying not to break them. Meanwhile NHM had decided to move out of the fraternity at the end of the spring term and began the search for an apartment and a roommate who would like to share it with him. For some time he had felt that there was neither an intellectual nor social environment in the fraternity that was of any value to him, but the final straw came one evening. A fraternity brother came into NHM's room the night before a difficult term paper was due. NHM was slaving away at the time to get his paper finished. The brother asked in a rather casual way, when is the paper in that class due. Tomorrow NHM said. Are you kidding, he responded? No why, NHM asked him? I haven't done mine yet, he responded as he left the room. NHM had no idea his fraternity brother was in the course. He must not have gone to lectures. A few minutes later the brother returned with a term paper in his hand. Is this an appropriate topic for the course he asked, as he showed NHM an old paper written by someone long ago. It was in fact yellowed with age. NHM looked at it and said, yes it is where did you get that? In the fraternity files he said and he left again. He returned yet a third time to show NHM that he had typed a new cover page with his name on it, and a slightly different title than the original had, and stapled it to the remainder of the old pages. The cover sheet was stark white, the rest yellow. Could he possibly submit that NHM thought without it being immediately obvious what he had done? Yes he did submit it, and proudly showed NHM the grade he had gotten. It was quite a bit better than NHM's grade. Seeing that he laughed at his success. Bye bye fraternity.

But then again, I might drop dead before I get to move out NHM thought. The results of the EKG showed however that NHM had an insignificant heart-rhythm irregularity that did not block his entry into the army. Once he regained his confidence, he signed all the military agreements. He called his parents and told then what he had done. They seemed not to be too concerned. NHM's mother was focused on the fact that her mother had become ill, that she might have cancer. His father offered some of his brass from the time of his active duty. He said that he wouldn't need it any more and that you might be able to use it. Thanks NHM said, I will, and when on active dutry he actually did! NHM regained his ability to walk normally, forged ahead in the search for an apartment, found an excellent one and a great roommate and began his new life as a research practice student not a farm practice student at Cornell as soon as the spring term was over. Two years were done, and he was still hanging in there.

NHM felt like a new man when the fall term began. Then one day in a zoology class laboratory of all places temptation overcame his better judgement. He simply could not resist taking a sheep's eye from the lab. The entire eyeball was intact. It looked repulsive. He knew it would be a useful item. So, when he went to select his specimen from a large jar of formaldehyde soaked eyeballs he took two instead of one and hid one of them in his briefcase. Later that afternoon when he got back to the apartment that he shared with his roommate and good friend Ken, he took it out and examined it to make sure it had survived the journey. It was in pristine condition. About the size of a large egg the eyeball stared out into space. NHM decided the best thing he could do to startle his roommate with it was to wrap it in crumpled newspaper and put it on his desk as if some trash has been put there. Ken was a student in the College of Engineering, the kind of person who kept everything perfectly neat. Trash on his desk would be an outrage to him.

There it sat until he arrived. Ken spotted it instantly. How did this trash get on my desk he shouted? Then he grabbed it to throw it in the garbage and the eyeball rolled out. He took one look at it and fainted dead away on the floor. This, NHM hadn't anticipated. NHM tried to arouse him but it wasn't effective so he dragged him to a nearby easy chair propped him into it and began to fan him. Just then their neighbor, a student who lived across the hall in another apartment in the big old mansion where they lived arrived and noticed the door to the hallway of NHM's apartment was open. Through it he could see NHM standing over Ken and trying to awaken him. The neighbor sensed something was wrong and he quickly entered their place. He saw Ken out cold on the chair, and then he spied the eye on the desk. Ken's eye must have come out he thought and he too fainted precisely as Ken had done five minutes earlier. What more could have gone wrong with both of them out cold? NHM started shouting for his neighbor's roommate to come and help but there was no reply. Fortunately Ken regained consciousness and then spotted Mike on the floor. What happened, Ken asked? Never mind NHM said give me a hand with Mike. They dragged Mike into the chair Ken had been in and finally he too awakened. They were both madder than hell but nevertheless wanted to see the eye again. Where did you get it they asked? What are you going to do with it they wanted to know? I'm not sure NHM said, I'll tell you tomorrow what happens.

NHM must have been out of control the next day. The brother of one of his victims told him afterwards that had the cops gotten there to catch NHM he would have been expelled. What happened was this. The two guys involved were two of NHM's new closest friends. Having left the fraternity NHM had become friendly with a number of graduate students who worked in the Plant Sciences Building. As a major in the Botany Department NHM managed to get some office space in rooms occupied primarily by the graduate students. There he got to know four people quite well: one named Dave Thompson was a genetics plant-breeding student. He had been the teaching assistant in NHM's basic genetics course and later became the CEO of his family's business the Ferry Morse Seed Company. His older brother Keith was an applied math/biometrics graduate student. Ken Sanderson was a Canadian who went on to became a well-known bacterial geneticist. Finally there was an Israeli, David Gershon who ended up working on animal viruses at the Technion in Haifa. On the day NHM had the eye with him, Dave and his brother had arranged to meet NHM for lunch in the home economics cafeteria on the Agriculture Quadrangle. It was the kind of place where you selected your food from various counters then took it to a cashier for payment. NHM took a few things, basically a drink and two slices of dry toast. Dave was just ahead of him on the checkout line. Keith had arrived earlier and was already seated eating a sandwich. While on line NHM took the eye out of his pocket and placed it upon the toast. From a distance it looked like a large hard-boiled egg. He pushed the tray along as if all were normal. The girl rang up Dave's lunch. He paid and moved on. She then started on NHM's with the usual smile and small talk then saw the eye. She immediately jumped up started screaming and ran out of the building. Everyone in the place looked to see what had happened. Dave turned to NHM and asked what happened? I don't know NHM said. They then walked to where Keith was eating and NHM sat down. Dave began to sit down too but he then saw the eye and became ill. He started to retch and ran out. Keith, calm and collected as usual muttered to NHM, yikes Neil, you best get out too. So, he did. He ran to his office, destroyed the eye and flushed it down the toilet at the end of the hall. For a moment he thought of leaving it in the toilet bowl for whoever came along but thought it too risky. The next day Keith told NHM the cops had come soon after he left to search for the guy with the eye but no one remembered what he looked like. Fortunately it ended there.

The courses that NHM needed to complete as a Botany major were mostly finished, but there was one yet to be taken that he dreaded. It was plant taxonomy. NHM knew that he really did not know one plant from another and that it was going to be very hard to learn how to tell them apart. There weren't many plants where he was raised, and later in high school it appeared that the tree pollens contributed to his asthma, hence he avoided trees as much as possible. NHM's inability to identify plants and lack of knowledge of their names was twice before revealed in the context of courses he had taken. The first of these was in a course dealing with the management of small forests that he took as an elective because his father had purchased thirty five acres of forested land in upstate New York without any idea of what to do with it or the trees that were on it. Although everything made sense in the lectures when they got out into the field in this woodland management course NHM was totally out of his element. In groups of two they were assigned to manage a section of the Cornell Forest Plantation. Among other things each pair of students had to thin the forest in their sector. That meant they had to cut down some very large trees. NHM was very hesitant to do that. What if I make a mistake and cut down the wrong tree he thought? These were not something that could easily be replaced. NHM insisted that the professor agree to every tree that they had to remove. The professor suspected that NHM had some problems and his suspicion was strongly confirmed by NHM's performance on the first tree-identification exam. The entire class went into the forest with the professor. The students were required to identify about twenty trees that he had selected and marked with numbers. NHM went to the first one and looked it over. The wind set it's leaves into motion and he quickly decided it must be a trembling aspen. Those leaves moved just as the professor had said they would in class. Then NHM went on to the next tree. This one was easy, looked like a Christmas tree but larger. Must have been a pine tree. The next in line got NHM worried. It too looked like what he imagined trembling aspen would, and he wrote trembling aspen for the second time. NHM nearly dropped when the following tree also appeared to be the same! He knew that was impossible. He knew he was sunk. So did the professor when he saw that NHM had gotten only four out of the twenty correct. The other students might have missed one or two at the most. The professor called NHM in to ask why he had done so poorly and he told him. Well here's what you will have to do then, he said. Go and visit the same twenty trees, and determine what they are using field guides. That was a hard task but NHM managed to get it done and saved himself.

NHM's ignorance of which plant is which was revealed to him in yet another setting. In this case it wasn't such a vital part of the course and he was able to cover up his deficiency almost forever. It was a wonderful and famous course in plant pathology. The main work of the course consisted of self-directed laboratories in which diseased specimens were supplied as well as the tools needed to determine what it was that was responsible for their pathology. At the first lab meeting the professor took students in groups of six to a greenhouse where he had assembled a variety of diseased plants. They all stood around the table with him and he asked each of them to tell him what they thought might be the matter with this or that plant. When it came NHM's turn he said, "Mr. Mendelson what do you think might be the matter with that tomato plant?" On the table in front of them there were about a dozen different plants. NHM had no idea which one was the tomato plant. He tried to figure it out however by observing which plant the others were looking at. Then he tried to think of what might have caused it to droop down and turn brown. Whatever NHM said led to a discussion and no one detected that NHM didn't even know which plant was the tomato plant. Many years later a colleague of NHM's moved from that same department to Plant Pathology at Arizona. By then NHM was a well-known scientist. NHM told him this story and he carried it back to the department and even to the man, still alive, who had been the professor. When he got back to Arizona he told NHM that his story made the rounds in the plant pathology community and was thought to be one of the funniest things they had ever heard. It definitely wasn't funny to NHM at the time. Nor was it helpful when he had to face the required plant taxonomy course. As anticipated, that course was the biggest hurdle NHM had to get over in his career as a Botany major. He sweated it out. In spite of the pressure he found some humor there as well. On field trips the taxonomy professor raced through forest and field to find examples of this or that plant. The students all ran along behind him to keep up. When he spotted a specimen he wished to tell them about he straddled it and shouted out, right here folks it's a new species. Everyone hurried to get a glimpse of whatever it was before he was off to find the next one. On one such trip NHM spent too much time looking over one of the finds, and thus ended up at the very back of the group when making their way to the next plant. The taxonomist shouted out his usual song, right here between my legs is a new species he said, to which the student behind NHM added, "Penis erectus". That did not please the taxonomist, but NHM loved it. He had to cough to hide his laughter.

Well into the third year of college NHM got a phone call from his mother with chilling news. First she said that his uncle had been murdered. Strangled by someone in what appeared to be a take-over move on the family business. At the time NHM's father was in the hospital being treated for an intractable bacterial infection. When they got the news he dispatched her to go and look at the corpse to confirm that it was indeed correctly identified. She insisted on telling NHM how gory it was. She said she wanted him to know all this because it was on the local news and she didn't want him to see it first on television. Never mind that NHM had no TV.

NHM appreciated getting the news from her. Clearly she was out of her element in having to deal with the horror by herself and although there was little he could do from afar NHM thought she was relieved by just telling him all about it. Then in a kind of matter of fact way she added that her mother indeed had a life-threatening cancer that hadn't responded to either surgery or chemotherapy. Her mother was essentially NHM's surrogate mother and the news struck hard. Is there anything I can do, NHM asked? No, nothing she said and their conversation was over. NHM went to the library the next day to begin a literature search in the hope that perhaps he could find some potential way to save his grandmother's life, perhaps an experimental therapy. Everything always came to the same dead end. When the disease reaches stage X there is no way of reversing it. Death was certain. NHM thought a lot about that. This was a cellular disease he reasoned. Given knowledge of cellular genetics why shouldn't it be possible to do something? I ought to be able to find some way NHM thought, but obviously he couldn't. It's too bad medicine is so primitive he thought and NHM felt very guilty about not being able to do a damn thing. He realized it would take a miracle for her to survive. What would it take for a miracle he wondered? On the streets in Brooklyn NHM had often heard the desperate say, pray for him or pray for her. Ridiculous he concluded, but perhaps worth a try. Then NHM realized that he had no idea of how one might actually do that. Perhaps I should look into it he thought. Could you learn how by going to religious services he questioned? Maybe, so off he went to the Friday night service at the Cornell United Religious Works program.

After a month or so the Rabbi noticed his presence and realized that he didn't know who he was. The Rabbi introduced himself to NHM and welcomed him to the "congregation". NHM said nothing of his goal but instead told him that although he was Jewish he had no idea of what that really meant and had decided to come to find out. NHM thought the Rabbi had probably never heard anything like that before but he wasn't put off by the comment. He replied by thanking NHM for coming and said that he would be happy to help answer any questions that he had. NHM thanked him and continued to attend but learned very little beyond the ritual and periodic readings of scripture. The issue of how to pray for a miracle never really came up. When his grandmother died within a year, the Rabbi offered his condolences; they said the Mourner's Kaddish and life went on. In the month's that followed the Rabbi asked if NHM would be willing to read sections of scripture in English at services each month, which he agreed to do. When NHM met his future wife, a young student working in a nearby laboratory, he asked her if she would like to go to one of the Friday night services. He did not tell her that he would be reading from scripture that night. She was surprised to say the least. She knew NHM was not a religious person, so what was going on she wondered? The Rabbi spotted her right away and quickly invited the two of them to dinner at his home. They both liked him and thought that perhaps someday he might marry them. It didn't exactly work out that way. Instead, the following summer a Rabbi at Fort Bragg in North Carolina married them. Strange as it all sounds it definitely worked. They celebrated their forty seventh anniversary in 2006! In 1959 when NHM told his friend Ken that he had met the girl he would marry, Ken said, "yeah right" as if that was just a dream. Now all he can say is, how could I have been so wrong?

After he finished his last semester at Cornell NHM skipped the graduation because of an obligation. He had to complete his basic military training before receiving a Commission. Off he went to Fort Bragg and his fiancée, Joan, went to New York City to finish a research project she had worked on earlier at the American Museum of Natural History and to plan their wedding. Their objective was to be married as soon as NHM completed the course at Ft. Bragg and then to leave for Indiana University where he would enter graduate school and she would finish her undergraduate degree. Joan found it impossible to get both families to agree to any of the details of the wedding and so she and NHM decided that they would instead have a simple wedding at Ft. Bragg after the commissioning ceremony. NHM arranged for the post's Jewish Chaplain to perform the wedding and they invited their immediate families to attend if they wished to. The families were not pleased to say the least but they had no options. Both families drove to North Carolina to witness an eighteen-year-old student/scholar marry a twenty one-year-old Ivy League Graduate and Commissioned Officer in the US Army. An unexpected storm blew out power to the Chapel forcing them to go to the only other place with power, the Officer's Club. It rained like hell until the next morning, fortunately letting up just in time for NHM to complete the second part of the commissioning ceremony, the part where they actually pin on the rank insignia. As soon as it was over, he hurried back to the barracks he had occupied and called for the Sergeant to come outside. "Get out here Sergeant", he shouted, "and make it fast." The Sergeant appeared as ordered, saw NHM as a newly minted officer and snapped off his best salute. NHM thanked him for his efforts during the training course and keeping them all alive. The Sergeant wished him good luck in his career. NHM hurried back to pick up his wife and off they went. Next stop: Bloomington, Indiana. Neither of them had a clue what it might be like, nor a worry. How could either of them fail? No, everything was under control. Talk about the confidence of youth.

When they drove into Bloomington, the home of Indiana University NHM's focus had so changed from living to become a farmer to wanting to become a scientist that he hardly noticed what mid-western farms were all about. It was of course agriculture on a very grand scale. Flat land with miles upon miles of corn or whatever stretched in all directions. It was green and beautiful and the air was at least as clear as it had been in Ithaca. But it didn't look as if NHM would ever get to farming. He was too far behind the curve so to speak in all respects to be able to accomplish that. Cornell had convinced him that he couldn't do the farm work as they deemed it had to be done. On top of that NHM realized that even if he were proficient at it he didn't have the money needed to own and run a farm. At Cornell NHM came to recognize that there were some things he could do competitively, things not so dependent upon the pieces that were missing from his education. These were the things he hoped to accomplish at Indiana. Science was his fallback position. He had to make it work at Indiana for there was no third default position. He knew he could never go back to New York City but what wasn't clear yet was where he could go. Neither Joan nor he really thought much about that at the time. Rather they had the firm belief that if they learned enough it would be enough and they would have enough. At the moment, though, they had very little.

The Indiana University campus was indeed beautiful though not as spectacular as the Cornell campus. They got a very warm reception from the faculty and other graduate students. Aside from the problem of finding a place to live, they settled into the academic community quickly. The contrast between Indiana and Cornell, between Bloomington and Ithaca, and between the students attending each institution was immediately evident. The pace of life in Bloomington was so much slower than that they were accustomed to that at first they thought time might be going backwards in Indiana. The perceived time warp turned out to be a reflection of the fact that Indiana University is a scholarly institution. The emphasis there was not so much on getting more facts than one could digest or understand, or to outdo everyone else including your professor, but rather on thinking something through, on learning and figuring things out, on exploring. The professors were on your side. They, too, were willing to learn something even if it was from a student. Neither Joan nor NHM had any idea that such an environment existed at a university. It was as if they had found gold. The cultures of the two institutions were as different as they could be. There was very little pretension at Indiana. When NHM wore one of his suits to class the first week a professor asked him if he was getting married or taking his qualifying exam that day. No, NHM said I'm already married and a qualifying exam is perhaps two years in the future. Where did you do your undergraduate work the professor asked? Cornell, he said. "Makes sense" the professor responded.. Things are much more casual here, he said; I went to Harvard it was the same thing there long ago. Well, the former Cornellian's reset their clocks and settled in for the long ride to a Ph.D., for NHM and B.A. and M.A. for Joan.

Joan in fact proved how good the quality of the education you could get at Indiana was by competing for and winning at a national level every kind of academic award you could imagine: a Ford Foundation Fellowship, A National Science Foundation Fellowship, and on and on. NHM managed to get a National Institutes of Heath Training Grant Fellowship and then one of the first National Institutes of Health pre-doctoral fellowships. The new NIH fellowship included funds for the purchase of some equipment. NHM told the professor in charge of his graduate research that he wanted to purchase a set of his own glass pipettes with some of the funds and asked if there was anything else the lab needed. His professor, Dr. Dean Fraser asked if there was enough money to also purchase an oxygen sensor electrode with associated electronics. Indeed there was. Then he made a suggestion that taught NHM something about the way grant money is administered at universities. What he said was, spend the money as fast as possible. He explained that the university hadn't yet had time to establish a policy concerning the use of funds for this award because it was a brand new NIH program. In fact NHM's personal notice of award came before anything the university had gotten about it. Dr. Fraser established an account in the department through which NHM could purchase against the award and NHM quickly bought the two main items. Together they nearly exhausted the entire budget.

The pipette manufacturer asked if NHM wished to have them engraved, no extra charge they said. Sure, I would, please have each engraved with NHM he requested. The pipettes arrived. They were a big hit and a very nice announcement to all that NHM had gotten the first such award at Indiana. Later when NHM noticed that some of the pipettes were disappearing in the dishwashing facility he posted the following memo: Attention. Please look to see if by mistake you have gotten any pipettes engraved with WHN, Wilbur Herbert Norton. If you do please let me have them back. They are part of my NIH equipment. Thanks. NHM, Neil H. Mendelson. If you turned the pipettes upside down NHM noticed they read WHN! Next came the oxygen electrode. It was very useful in the production of large batches of bacterial viruses of the kind they worked with in Dr. Fraser's lab. The following month NHM got a memo from the department head. It said that the University had established a policy concerning the NIH predoctoral fellowship equipment allowance. NHM would be allotted forty percent of the amount awarded. NHM went to see him and told him he had already spent ninety five percent of the funds and the equipment was in full use. NHM said he had no idea the university would take sixty percent of the equipment budget. What about the salary and other budget lines of the award he asked him? They won't be reduced the department head assured him. Thanks he said and that was the end of their discussion.

When NHM finished his Ph.D. most of the pipettes were still intact. The laboratory inherited them for use of the next generation students. The oxygen sensor remained in use for many years afterwards.

At Indiana NHM learned that there were unfortunately some sharks lurking under the calm waters. Isn't that always the case? When you least expect it, boom, you're a victim. In the attack that he witnessed the poor victim was critically injured at the very end of his Ph.D. program, denied any degree, and had to leave for an academic job without any credentials. NHM never did learn what his ultimate career was like. This was the second such failure that NHM had experienced in his short academic career and it made him think that nothing is really secure in this profession. Little did he know at the time how pervasive that insecurity was at all levels in the academic community. It was 1961, his second year in graduate school. NHM was working in the laboratory of a well-respected cellular geneticist, a man who had done some wonderful experiments and who was as charming a person as one could imagine. On Friday evenings he hosted an informal meeting at his home to which graduate students and some others were invited, where a topic of interest to his research program was discussed. A member of his research group was always asked beforehand to lead the evening's discussion. The background and all the facts were given by the leader to start things off, and then others explored the issue in detail, raised questions, or made criticisms of either the technique used or the conclusions reached. As a student in the laboratory NHM was always invited, and from time to time asked to serve as leader. At first he found these gatherings quite useful and learned a great deal from them. Then everything seemed to stagnate as if all the ideas had already been posited. The group had run dry of anything new to offer. The host detected this as well and announced one evening that he had something different planned for the following semester. He told them that he had asked a senior graduate student who was working on a problem related to the interests of his group to come and tell them about his thesis research. It promised to be a rewarding semester and indeed it started well. About a third of the way through the thesis experiments the host began to strongly criticize the findings. In each instance he asked for proof that the phenomenon described was a genetic character rather than a physiological one. Although the student showed quite well that the phenotypes involved passed from parent to progeny, he could neither identify a gene responsible for the traits studied nor guarantee that they would persist permanently in the progeny. Neither NHM, nor others attending could really understand the objections being raised. Particularly so because the host himself had in another experimental organism shown quite similar patterns of "inheritance" and made a case for their being a new kind of "genetic" information residing in cell structure rather than in classical genes. Shortly after all of his thesis experiments had been scrutinized at the Friday night meetings a Ph.D. committee meeting was called for the student who had provided all his data to the group. NHM was sickened to hear that his thesis research had been deemed unacceptable and he was asked to leave the program. For what reason did they terminate him? The shark had attacked and forced the committee to dismiss the student. The student's major professor was furious. Another famous geneticist was also a member of the rejected student's committee. This man was then the head of the department in which the failed student had been working. The department head immediately established a policy prohibiting any students in his department from having the shark on their Ph.D. committees. But it was too late for the victim. It is never safe to swim in the same waters as the shark, NHM thought. I shall have to find another group to work in as soon as possible, he decided. By that time NHM's abilities in genetics and related sciences were well known by the IU faculty. He thought it likely that he would be able to move without difficulty. NHM decided to move into a molecular biology laboratory where he would be able to work in an area of genetics that was then at its zenith, the study of gene function in bacterial viruses. NHM was happy to be given a place without any problems.

In the move, he discovered by chance a shark attack targeting him by a professor at Cornell whom he had counted as one of his strong supporters. This chance revelation really opened his eyes. He learned of it while transporting his graduate file from his previous laboratory to his new home. NHM had neither any idea of what might be in such a file nor any inkling that it might be confidential. The secretary from the genetics program called him and asked if he could deliver his file to the new department, so he picked it up and took it to the new office. The office was closed for some reason when he got there so he stopped by his new lab to find out when it would open. The people in the lab told him it would only be a matter of five or so minutes. NHM sat down at the lab bench to wait and opened his file to see what it contained. Among other things it had several letters of recommendation sent to IU by supporters from Cornell. The first one he read ruined his whole day! It was the poorest letter he could have imagined, suggesting at its conclusion that NHM should only be accepted into the IU program if they had an extra position they simply had to fill that year. No reasons for his opinion were given. Why on earth would they have accepted me after getting such a recommendation NHM wondered? Why did he bother to write if he thought I shouldn't have been accepted? The author was an excellent scientist who gave one of the best courses in the Cornell program. NHM took it and did well, and sat in on it again for a second time to hear about the newest things. He couldn't believe that this seemingly friendly Professor was a shark or what his motivations might have been to attack NHM. Rather than delivering the file after making that discovery NHM took it to his office and waited for his wife to arrive after her class. She knew the professor quite well. She too could not understand what might have brought on his ire.

That was in 1961. By 1966 NHM was a National Science Foundation Postdoctoral Fellow working in London, having survived two years in the army during the Vietnam War. While in London another postdoc arrived several years his senior who had done his Ph.D. at Cornell. Guess what, he too knew the character of the Cornell shark. Those closely associated with him were well aware of the problem he told NHM, but no one ever said a word about it to anyone else. They both laughed at having escaped his jaws unscathed. Jump ahead another eight years. NHM had organized an international meeting for the American Society for Microbiology that they had decided would be held at Cornell. NHM was on the campus for a week and took time off to see which of his old professors were still alive. The shark was one of them. NHM made an appointment to see him. By that time NHM was well known for many discoveries. He did not let on that he knew of the shark's letter. The shark was as polite as ever. NHM asked him if he knew about his work. He did not. Nor did he know what the cutting edge of work at the time involved. The shark told NHM that his course had become a history course. He was locked in the 1940-60 era. Did he in fact remember the letter he wrote for NHM? He definitely knew who NHM was and discussed events of the days when he was a student there. NHM's visit shed no further light on his venomous deed save to reveal that his façade remained intact through to his old age. Who knows what further damage he might have done to others over the decades. Achievement is the only way NHM knows to defeat these marauding predators, but you have to survive the initial attack to get your chance and that is not always possible.

There was much to be learned at Indiana not only about science and research but also about people and culture, the frustrations of trying to do something really unique, to discover something really new and the problems of making the transition from consumer of information to producer of information. Unless that transition is made, a person never really becomes a scientist. They might get by as a perfectly respectable farmer or teacher or physician or engineer or administrator but they can't be a scientist. Indiana is the place where NHM gained the confidence needed to become a research scientist. He got to know many very successful scientists and he couldn't see any reason that he shouldn't be able to pursue research the way they did. NHM had no difficulty synthesizing the facts, understanding the logic or the experimental approaches then used, and he felt comfortable reaching conclusions given data. NHM knew his analytical abilities were on a par with those of his successful seniors by challenging them in many ways, and they were receptive to the test. When he beat a very distinguished visiting scientist in solving a problem that appeared in the Mathematical Games of Scientific American involving operations research, something about which neither of them knew anything, NHM's confidence grew. When he challenged published results in leading papers by pointing out either errors in logic or failed controls his criticisms were upheld by the expert professors in the field. He became bolder yet. NHM's stature among the graduate students rose and he became inclined to do some crazy, outlandish things that could have rapidly sunk his ship but fortunately didn't.

NHM will never forget one such dirty deed. It had to do with a lecture that every doctoral candidate had to give to the faculty and students of the program, in his case, the Graduate Genetics Program. He was aware of three newly discovered systems of inheritance that stood in clear violation of Mendel's classical laws of genetics and chose to discuss them in his lecture. At that time there were no mechanistic interpretations of the observed phenomena. NHM read everything that had been published about these systems. In one of the papers he discovered a critical missing control experiment that raised serious doubts about the validity of the conclusions. In the other two systems he found ways to explain the results on the basis of molecular mechanisms that were only then being worked out in detail. He knew the talk would be of interest to a lot of people, that there would be a large audience and that he would enjoy demonstrating what he could do. Just a short time before the assigned lecture date NHM heard about an object that one of the graduate students in zoology had made. It looked much like a walking stick but in fact was a bull penis stretched and stiffened by the insertion of a metal rod, preserved, and given a glossy shellac finish. NHM went to see the student who made the object and asked to see it. Sure Neil, he said, here it is. It was as good as had been rumored. Then NHM sprung his suggestion upon him. Would you lend it to me for a day he asked him? Here is my plan. I would like to use it as a pointer for my graduate-qualifying lecture next week, NHM told him. "Would you really do that, he asked?" "Yes" NHM said, "if you lend it to me I would do it." "OK, take it", he said, "Wait, I want to check that my wife has no objections first". "Go ahead take it," he repeated, "if she objects I'll come and get it back from you." "Great", NHM said and he quickly scurried away with the surprise he was planning for his talk.

NHM's wife Joan was not sure he ought to use it but he finally convinced her that nothing really bad could happen and she eventually went along with the prank. When NHM got to the lecture room it was already quite full. He took all his gear up to the front table, organized the data charts then sat down waiting to be introduced by the professor in charge of the Ph.D. qualifying talks. His lecture went along as he had hoped it would. Various professors asked questions from time to time but NHM tried to keep things going according to schedule. The audience appeared to be captivated by the material as NHM assumed they would be. And they were enthusiastic as well. From time to time he used the penis pointer but nobody appeared to notice it. Toward the end of the talk NHM realized that he couldn't get everything covered that he had planned. He summarized everything that he had talked about and mentioned that he would be happy to explain to anyone interested afterwards what critical control experiment hadn't been done and why that cast doubt on the main conclusion for the third system. Before he knew it the talk was over. About a third of the audience left after the formal question period concluded. The remainder hung around to ask other things or listen to the discussion that ensued. They all gathered around the front table where NHM had all his materials. One of the advanced graduate students working in developmental genetics was among them. So too were a number of the senior professors in genetics. The developmental genetics graduate student, a young woman about to finish her Ph.D. in the lab of a man who had invented the method of nuclear transplantation in frogs, was very interested in what NHM had to say. She had recently discovered that many chromosomes in transplanted nuclei were unexpectedly damaged and thus years of research conclusions based upon nuclear transplantation work were now placed in question. In the course of NHM's discussion with her she noticed the pointer, picked it up, examined it and being unable to identify it asked him, Neil, what is this? Everyone in the crowd became silent. Take a good look at it, he said in response. Don't your recognize it? No, what is it she continued as she examined it again very carefully? That is a genuine bull penis he said. The look on her face should have been captured on film. She was repulsed and terribly embarrassed, quickly put it down on the table and backed away as if it were infectious material. At that moment NHM happened to notice the face of Dr. Marcus Rhoades standing directly behind her. Rhoades was the head of the Botany department, a very accomplished and famous cytogeneticist, and a member of NHM's Ph.D. committee. An older man, he saw all the humor of NHM's nasty trick. The rest of the crowd roared with laughter. Even the victimized woman began to laugh. After all she was in the Zoology Department and certainly had dealt with much worse things. The following week the head of the Zoology

Department, Dr. Theodore Torrey, called NHM in. He was the prototype starched-shirt conservative of the old formal school but very honest and kindly. In view of NHM's unusual arrangement of belonging to one department and working in another, Torrey served as the administrator of NHM's Ph.D. committee, as he had done once before, in the case of Jim Watson, the man who with Francis Crick had discovered the structure of DNA. Dr. Torrey told NHM he had heard of the prank and thought it in bad taste. NHM agreed with him. Torrey suggested that NHM apologize to the offended party. NHM told him that he already had done so and she had accepted it. Good he said, try not to do anything like that again please. I will NHM said and off he went. Another close call he thought on the way back to his lab. NHM felt proud that in their meeting neither Dr. Torrey nor he had burst into laughter. It was certainly just below the surface in both of them however.

During their days at Indiana it was indeed a financial struggle for Joan and Neil. Initally Joan had given up a New York State Scholarship in transferring from Cornell to Indiana and NHM lost his income from playing in the band he had formed in Ithaca. Their first term in Bloomington brought in one hundred and sixty two dollars and fifty cents per month from NHM's teaching assistantship. That had to cover everything. It barely did. NHM found an unexpected source of some additional funds. One of the students in a laboratory that he taught mentioned to him that he had returned to school after being in the Army. When NHM told him that he was a commissioned officer granted a delay from active duty until he finished his Ph.D., the student asked him why he didn't join the reserve unit in Bloomington? He told NHM that he was a Captain in it, that it met two evenings and one weekend day a month. In addition there was a two-week summer camp. The meeting schedule was set up just for the convenience of IU personnel. He said most of the officers were affiliated in one-way or another with IU.There were many benefits. He said come to the next meeting and see what it is all about. NHM did and he joined the unit. He attended regularly, did what he was assigned to do and went to camp in the summer. The extra money it brought in was definitely a help. In a short time NHM was promoted to first Lieutenant, and his pay went up. At the weekly meetings NHM got to know several people in the community that he would never have met in the normal course of things. The first and most significant one was a black officer who was doing a Ph.D. in sociology, Joseph Walter Scott. Joe learned that NHM was a student of H. J. Muller's, the Nobel Prize winning geneticist and wanted to know as much as he could about the nature/nuture issue, in view of the fact that some bigots had stigmatized blacks as an intellectually inferior race. In their discussions it became clear that Joe knew a great deal more about genetics than NHM did about sociology in spite of his life on the streets of New York City. To compensate for this deficit NHM decided to attend a course Joe taught in the evenings. From it NHM got a new perspective of the way in which the structure of society prevented the blacks from achieving whatever potential they had. The friendship between NHM and Joe Scott has lasted to this day.

NHM never fails to learn something from him when they get together; sometimes even when they don't get together. Here is a perfect example. While sorting out old papers in preparation for the writing of this book NHM came upon a paper Joe published in 1968. It's title is, "A Perspective on Middle-Class Delinquency". There were two authors, Joseph W. Scott from Indiana University and Edmund W. Vaz from McMaster University. NHM thought Joe might like to know that he still had this old paper of his and decided upon a unique way of telling him. NHM phoned Joe one evening and got his answering machine as usual. Joe liked to screen all his calls this way. So, disguising his voice, NHM said, "Joseph Scott this is professor Edmund Vaz calling. I will try to reach you at another number." That's all, no call back number, no mention of what it was about or anything. NHM left the same message on Joe's other phone. He assumed Joe would figure out who might have left the message and would call him back. But no call came for weeks. Joe might be away between teaching obligations NHM thought and just kept waiting. Finally a call came from him, but it had nothing to do with the message NHM had left. After discussing what Joe had called about NHM said to him, "Listen Joe, I have an old paper of yours" and he read the title and authors to him. Joe shot back, "Oh no, Neil that was you who left the messages wasn't it." "Yes it was", NHM replied. You won't believe this but listen. I haven't heard from Vaz in over thirty years so I tracked him down on the Internet through his books. He's eighty years old now and in a nursing home in Waterloo. His wife died three months ago, then he fell down and broke his hip and landed in the nursing home until it heals. He was thrilled to hear from me. Vaz asked me why I had called him. I got your phone calls Joe told him. I didn't call you Joe, Vaz said. I thought he had Alzheimer's and forgotten Joe said, so I went ahead with the conversation anyway. I brought him up to date about mutual friends and let him know what had gone on for the past twenty years. Vaz loved it. Joe, you should call him back and tell him what happened, NHM suggested. No, I think I will let well enough alone, Joe said, you never know what comes of your jokes Neil. In this case it was a very good thing.

In the reserve unit neither Joe nor NHM had much time to discuss anything. They were among the lowest ranking officers and although NHM could get all his work done in a reasonable time; Joe was always given more to do than anyone could possibly finish in the assigned time. They really piled it on him. Then, one day, the commanding officer summoned NHM to his office. He was the backfield coach of the Indiana University football team and the owner of a very large slum in Bloomington. Lieutenant Mendelson, he said, I have seen that you have become friendly with lieutenant Scott. Yes I have NHM said, we are both Ph.D. candidates at IU. Yes he said I know Scott has the smarts but I want to warn you that he is lazy and no one likes him. You're a good officer and I don't want to see you disliked because of whom you hang around with in this unit. Now I can't order you to do anything about this but I just thought a word would be helpful. Yes sir, NHM responded, I understand and thank you for letting me know how you feel about this matter. Indeed NHM did understand, his commanding officer was simply prejudiced. Of course NHM told Joe about what the CO had to say. It wasn't news to him. Many blacks he knew in Bloomington lived in the slum that the CO owned. They knew his character very well. How strange NHM thought, this man's civilian job in the athletics department at IU must bring him into contact with many black athletes everyday. He must be very uncomfortable at work. Joe warned NHM that the CO was not the only officer in their unit who disliked blacks. His immediate commander, a man who had invited NHM to attend a party honoring his recent promotion to Captain, was among the most prejudiced of the lot. Haven't you noticed I get twice the work assignment that anyone else does, Joe asked? Yes, certainly I have NHM told him. Well, why do you think he does that? He makes sure I can't get it all done in time so he can write a poor evaluation of my work performance. These go to the unit's commanding officer, who puts into my record that I am a lazy and incompetent officer. It's a well known method, Joe said. We shall see who ends up better in the long run. Yes I hope we do, NHM responded. Neither of them knew then however that their time in the Bloomington Reserve Unit would soon be over. They didn't have the long run to declare the winner at least not there. Joe and NHM and several others were ordered out of the unit within the year because they had not yet attended their Army Branch School. Why did that happen?

The Vietnam War was beginning to escalate and there was a chance that reserve units would be activated. The Pentagon realized that there were officers throughout the country who had joined active reserve units while in pursuit of doctoral degrees of all sorts. These officers had been granted a delay of four years duration before being activated. Only then were they sent to their branch schools and to their final duty stations. Without having been branch qualified, these officers would not be able to go along with their unit if it were activated. The Pentagon decided therefore to replace all the ineligible officers with others who had already completed their military branch schooling. Orders were sent throughout the country requiring officers such as Joe and NHM to leave active reserve status. By that time NHM had been promoted to first lieutenant. At the University NHM had been made a fellow on the Genetics Program's NIH training grant. In addition, his wife had won a Ford Foundation Fellowship, so they were feeling rich. They saved twenty five dollars a month from their income. On top of all that, of all things, at just about that time NHM decided to display some paintings he had made as a hobby at a street fair in Bloomington (the first annual Kirkwood Arts Festival, as it was called) and two of his paintings were purchased. One was a small watercolor scene of boats in harbor; the other was an oil painting depicting the main street leading to the entry gate to IU. A woman whom NHM did not know bought the little watercolor. She said it would be given as a birthday present to her daughter. NHM later learned that she was the wife of a biochemistry professor he knew quite well, an older man from Vienna. At a meeting the following week that they both attended the professor overheard others teasing NHM about being a scientist/artist and told them in no uncertain terms that everyone in science should have some link to arts and humanities. He gave the fact that he was a pianist as an example. That quieted them down fast! The second painting was bartered for three pairs of slacks from the owner of the men's store that appeared in the painting. The owner hung it in his store. Occasionally NHM stopped in to take a look at it. NHM tried to reproduce it but he couldn't. Too bad I "sold" it he thought. The pants wore out but the painting is probably still in fine shape. The pants did serve however until NHM was ordered to active duty two and half years later.

Unfortunately Joe's poor ratings in the reserve unit delayed his promotion to first[t] lieutenant quickly enough so that he could enjoy the benefit of higher salary. Grants in support of graduate students in sociology were much harder to find than in genetics, so Joe could have used the extra money. He nevertheless pushed ahead and kept his family secure. When NHM's orders finally arrived instructing him to report to his Branch school he had just finished writing his Ph.D. thesis but didn't have time to arrange for the final defense of the thesis. The department granted him a delay until he could get back for the final formalities consisting of a lecture followed by a closed-door question session with the Ph.D. committee, and finally a celebratory party. Joe was ordered to active duty shortly after NHM was. He too had finished his graduate program and was now Dr. Scott. Nevertheless the army sent him to his Branch school, the Military Police Branch School in the deep south, still very segregated and a rather dangerous place for a black. NHM went to visit him in Georgia just before finishing his own schooling in Alabama. Joe and his family had been forced to live off the base where they were more than persona non-grata to say the least consequently he would not go out with NHM where the two of them could be seen together. It was indeed somewhat frightening. His "friendly" neighbors attempted at least once to burn down the house he rented. Still the army refused to assign him on-post housing. The Pentagon did relent however when it came time for his final duty assignment. Rather than make him a police commander, they gave him a research assignment where he could work on developing policy dealing with minority members of the armed services. NHM's final duty assignment came as no surprise. He too was asked to work in a research capacity rather than as a staff officer. Several months after he starting working on his assigned projects he had accumulated some leave time and requested a short leave to return to Indiana for his Ph.D. final defense.

Given life's situation, NHM's Ph.D. final defense seemed like nothing more than a trivial formality. All the requirements for the degree had been met and that chapter of his life was over. The official parchment arrived in the mail one Saturday at their house in the officer's section of the installation where he pursued his research. NHM remembers the event well. He and his wife were cleaning the house at the time. They propped up the Ph.D. parchment on the dining room table and NHM mopped underneath it while his wife cleaned the living room. They both enjoyed this ritual knowing full well that the job at hand was to survive the remaining eighteen months. Being "Doctor" by comparison was insignificant at least for then. Fortunately both Joe and NHM lived through their two war years. For that they are both grateful. NHM worked in the Chemical Corps. Years later it had to be discontinued as a result of political pressure following leaks that revealed some of the awful things that went on there. In Joe's case he became a highly respected officer for the work he did but resisted the temptation to extend his active duty tour in spite of recommendations that he be placed on a fast track leading to the rank of General. Instead he became a Professor and so outdid those back in the Indiana Reserve Unit who plagued him that there can be no question of who won in the long run. NHM is glad he lived to see it!

12. The heart of the matter, part II

Well now we have seen the factors that limited NHM's choices, drove his aspirations, and defined the kind of professional, and, to some extent, personal life that he was destined to lead. Escape was a major motivation. Self-reliance was an absolute necessity. Without any valid role models in his environment, a great deal of effort had to go into filling that which was absent in nearly all aspects of his life, including people, culture, professions, ethics, even the basic elements of survival. Whom to trust and whom not to, what was a reasonable goal hence worth attempting and what wasn't, where to go and where not to, what was safe and what was dangerous, how far can one get from zero? When you climb out of a deep hole those who knew you when you were at the bottom often think, oh, he had to do that to show that he was better than so and so, or he believed all that propaganda they fed him about how great he would be one day, or the like. NHM does not think any of that truly pertains to his situation. He believes his confidence came from getting through repetitive challenges unscathed, as Uncle Wiggly always did in the stories he remembers from his childhood. NHM wasn't stronger than anyone else, healthier than anyone else, better at anything than anyone else in the group, He just muddled his way along with very little planning, blown in one direction or the other at any given time, but he was not willing to give up easily. Survival was essential. Isn't it always? Having discovered his predicament at an early age he dreamed up a solution for the two major plagues that he thought he must escape: the inner city setting, and the people/culture that existed there. He was sure it was a great plan and NHM stuck with it as long as he could for as he grew older the idea that he would do best in a country setting and by working alone at whatever he did became more and more reinforced. Competition was not his thing. Obviously. How could he be competitive in anything given what he knew or could do? To win then he had to find a way to do different things, not what everyone else could do, but something they either hadn't thought of yet or didn't care about. Farming was the way out. His only way out.

It was a perfect solution but unfortunately doomed from the very start by two things NHM did not know then: that he is an asthmatic, and that no one can simply walk in and start a farm from nothing. Even to take over an existing farm requires a great deal of money as well as the knowledge of how to be a farmer. NHM might have been able to work around the health limitation but the financial barrier was insurmountable.

Could there have been another way to achieve his life objectives, a substitute perhaps that achieved the goals but in a slightly different way that might have been doable given his disabilities? Not that he could identify as a child desperate to get out of the scary situation he was in. No, all his creative energies appeared to have been spent on coming up with plan A, becoming a farmer. There was no backup or contingency plan. He knew what he needed and what he wanted and that was that. Pure chance pushed him through to another career. It wasn't quite a perfect match but he could live with it.

Later in life when NHM first realized that he was going to become a scientist he thought what a clever way to avoid competition. After all, scientists make discoveries, that is, they find things that no one else knows, and therefore there can't be any competition. Scientists can work alone in their own laboratories, kind of like being on a farm. Not only that, a scientist's laboratory doesn't have to be bought with his own money as a farm does so you can start with nothing. Never mind for the time being where the money comes from. How would anyone in NHM's position know that anyway? These were the thoughts that went through his mind when it became clear during his years at Cornell that his professors believed he too could be a professor and encouraged him to continue studies through the Ph.D. They thought he could do what they do. He thought what they did was what scientists did and that he might be able to achieve his core needs in life doing the same thing. If plan A was dead, then plan B, becoming a professor, might be a good substitute. In truth NHM had nothing else to turn to. The things he learned as an undergraduate were not exactly the foundations being sought by those who came to hire Ivy League graduates. With the exception of the band that he organized and played in there really wasn't anything else he could point to that might have been considered a valuable life skill by an employer. Planning ahead was clearly not NHM's strong point. He was more concerned with getting through one day at a time. Given all of that he couldn't think of anything better than plan B. Of course the image in his mind of plan B was about as far from reality as could be imagined but this is the way it always was for him. Like all other things he had to learn the truth by living through it. This is an awful way to have to make major decisions about events that set a course in life and that might not be reversible. Plan B might have been a perfect mistake. He knew that but given the dimensions of what was missing in him that was all he had to work with.

Imagine seeing just the tip of an iceberg and from that trying to figure out what the whole thing might be like. That's kind of the way NHM operated during the years when he had to acquire the basic understandings that guide decision-making. That being the case it must be a miracle that he had been able to succeed in both science and music. Yes, he did put relentless effort into both but that alone couldn't have done it. Yes he had very good luck, but the magnitude of what NHM has done doesn't fit a beginner's luck model very well. NHM is reminded of a similar situation that illustrates the way in which he dealt with acquiring information about how people lived and felt outside of the sphere of influence that constituted his world. It was when he taught himself to play the piano during his teenage years by learning the popular music of the 1930's forties and fifties. It was the music of Broadway and supper club performers dealing with life's sorrows and happinesses, the spectrum of the human condition. NHM drank it in as if it were the literal truth. Then he lived it down in reality. The truth always emerged as time went on. No, you didn't have to finance your own laboratory, but someone else did. Who did? It could have been your host institution, but that never came his way. Well if not that source then where? Well the government seemed to be in the business of providing money for research. Yes, but to get it you needed a reputation, supporters, achievement, others to say how much promise you had to make significant achievements. You had to compete against all others also seeking the limited amount of money available. Not quite an outsider's game as farming would have been. How could NHM surmount this? Luck must have played a big part to get him started. He was the one who found the first important DNA mutant in the group at the Microbial Genetics Research Unit in London. True, he dreamed up the way to find such mutants but that was only an idea, not a proven method. Yes he pushed his idea as hard as he could in the lab, but others there were also working away every day, hunting as he was for the important gene. Once he found the prized wonder, he pursued characterizing it as relentlessly as possible. He knew that a race was on, that time was not in his favor with but a year's worth of support for his stay in London. He had no job in the United States to return to, and assumed that his only hope of getting one was to maintain the advantage that he had by possessing a rare commodity, the mutant he discovered. In a way this was a true test of his idea that he could be competitive in science by working on the level playing field where no one knew anything about the newest thing. This mutant was his gift. It was up to him to build it into the foundation of his career. His mistake was in believing that the building process was nothing more than achieving a good thing, as if that was money in the bank so to speak that would accrue interest sufficient to keep him alive. The foundation that he built for himself was therefore not as robust as it could have been. It was definitely strong enough though to support the edifice he built upon it which over about a nine-year period led to new discovery and new head starts. These in turn added to the foundation. By the time NHM discovered helical growth and developed macrofibers from it in the 1970's he wasn't just some unknown person asking for money from the government to support a totally unheard of and strange thing of no clear cut significance to anything. No indeed. NHM was a person who knew nothing except what he found and/or taught himself. Others of course didn't know the first half but what they did know was adequate. His theory worked. There is a way to bypass an enormous missing part and to go forward.

NHM came to realize that he bore some responsibility for all that was missing in him. The void he lived with was to some extent a product of his own doing! That is an unfortunate ramification of having been confronted repeatedly in his childhood with the difference between what others told him and the truth. NHM ceased to hear the cacophony of their voices and thus lost the chance to benefit from the small amount of what was correct amid the distracting noise. This shell in place, this limitation, forced self-education upon him in almost everything. So little trust was given to what others said that he regularly had to "reinvent the wheel", thus drastically limiting what could be appreciated by taking the short cut of accepting what others already knew. The need to learn it alone meant he had to isolate himself and choose carefully what it was he wanted to know. He had to accept the limits of his own abilities:

there are some things he may never be able to teach himself. When he did figure something out, he become very confident that he really understood it, at least at his own level. If he got it wrong, which of course was possible, it took a lot to convince him of his errors because the way he had seen it might really be something that the others had overlooked. The mind is a wonderful thing to challenge. In science however, fantasy must be separated from reality and that is why it is so important to be able to test theory by experimentation. For someone in NHM's situation this means you need to have skill with your hands as well as your brains. The lucky thing in NHM's case is that he could both think and explore. The latter was not as simple as it might sound. That shell within which he existed meant he really couldn't just use the tools and approaches of others (with very few exceptions) so he had to invent the experimental approaches used to test the theories arising from his own discoveries. Why would anyone waste time on such a difficult individual? If it weren't for the need of money, NHM probably would have never been bothered by this question. He was somehow content with what he could do without needing others to also appreciate it. When they did, however, both their appreciation and that fact that he enjoyed their appreciation surprised him. Being linked to something must be better than being totally alone.

NHM's first significant link to something came when he found his wife. Although she was only seventeen at the time, she knew everything. She was his window to the world. He knew that what she had already accomplished he would never achieve. All the foundations were there and the rewards that go with having mastered them were too. When they first got to know each other NHM is sure it appeared to her as if he was the one who knew everything not the other way around. After all he was a senior, she was a freshman, he had survived it, she knew she would but hadn't yet. NHM did his best to keep her from learning the truth about what was missing in him. When she first met his family she immediately sensed there was something very wrong but the full extent of it and its history couldn't be deduced in the brief encounters they had. NHM saw in her family both the genetic and cultural reasons she was what she was. She did not hide from him her own difficult childhood, which he believed could be put to rest permanently by their marriage. Right from the beginning of their lives together NHM's deficiencies came into sharp focus. NHM knew he had to make good for the both of them in spite of his disadvantage. Neither of them really knew how he could. The friends they made as a couple saw immediately how polarized they were as a pair: she the core person, the one they could communicate with, NHM the wild man, the entertainer, "the God knows what might happen next" partner. "How do you live with him", was the most frequent thing she heard. You didn't do so badly yourself was what NHM heard. Indeed he didn't. Total knowledge, refined culture, smart as hell and a beauty on top of it. Too bad she had to waste so much time as the conduit of light for NHM's existence. Not only light but everything, including their financial survival. Her burden could not have been greater yet she bore it without malice. NHM knows that in some ways today she thinks of him as Superman given what he has done across a spectrum of things. (Not because on his sixty sixth birthday he rented a superman costume, dressed in it, hid in a closet until his grandsons found where he had disappeared to then sprung out with the words, I am not Neil, I am the man of steel. As the Superman version of NHM flew around the house he came upon his daughter, Debbie, who lives in Los Angeles and who, had flown in as a surprise. Could I be flying that fast he said as they met?) The truth is that it is Joan who is the Superwoman. Not a soul would deny that. NHM is the also-ran who ran up hill all the way.

The main thing NHM hasn't been able to explain to himself or anyone else is why he found it impossible to let sleeping dogs lie. Why couldn't he go on beyond failures or things never mastered in this or that rather than to try to rectify the failure to turn it into gold as if an alchemist? Why did NHM always feel obliged to pay back debts and demand justice in everything? Why did NHM always have to surmount what he believed to be his own inadequate performance? To do these things took an enormous part of his life's energy. His career in music is a good example. There was no music in his home as a child. He never heard the first piano he ever saw played by anyone, the one at his grandmother's sister's home in Brooklyn, yet he was drawn to it; perhaps as an object, or anything beyond his world. Even so, he wanted to play it. He never heard the sound of the double bass when it was offered to him in high school as an instrument to learn, but he seized the opportunity as if he knew it had a beautiful voice. He never heard a piece of classical music played before he discovered a room in the student union at Cornell that played recordings that people requested. NHM remembers looking over the list of requests there on one occasion. There wasn't a single composer's name he had ever heard of. No wonder he couldn't put anything down as a request. Yet NHM has become a classical musician, paid to play in churches and other venues, as well as the principal bass in several orchestras including the Southern Arizona Symphony Orchestra (SASO). Over a twenty nine-year career he has probably played some works by every composer of the baroque, classical and romantic periods and he hasn't been able to escape playing the difficult music of modern composers as well. For as long as Observer can remember NHM has spent at least an hour a day playing the bass beyond rehearsals and concerts. He really has had a second career working with musicians who had actually been educated at conservatories or music schools. Many of them make their livings as musicians, both performers and music teachers. How this came about is quite parallel to other things in NHM's life, largely a matter of chance.

The bugle led to the trumpet. Asthma ended that but led to the bass. And the lack of adequate training on it left NHM with a frustration that lasted forever. By chance the scientist next door to him in the Shantz building, Don Bourque, was a professional bass player as well as a professor of biochemistry. NHM told him he hadn't played in over 20 years. Don asked if NHM had a bass. NHM found it in his parent's attic when visiting them after attending a meeting on the east coast. When it arrived in Tucson, Don Bourque informed him that everything had changed since NHM put the bass aside. He gave NHM a set of the new metal strings and recommended the current methods book, which NHM purchased along with a cheap bow. Having gotten through about three-quarters of it on his own he left for a year in Paris where he attempted to find some music teachers for his daughters by going to seek help from the professor of bass at the National Conservatory of Music. Of course they wouldn't let NHM in to see him when he got there. A visiting professor at the Pasteur Institute carried no weight there. But he forced his way in nevertheless. There the music professor was in his master class with all his students and in bursts this guy in a cowboy hat. In French NHM asked for a moment and apologized for the inconvenience. The professor came into the hall to speak to him and he explained his mission: to find a cello and a viola teacher for his daughters and perhaps a student of his to teach NHM on the bass. The professor asked if he could call NHM at home that evening so he gave him his phone number. Much to his surprise when he called, he asked if NHM would be willing to study with him! John-Pierre Logerot. Holy cow how could that be? Within the week Logerot and his son appeared at NHM's apartment with a bass in tow, a bow and the name of the French method book they would use. At the end of the year he agreed to sell NHM the base that he had rented. It is a very fine French instrument made in about 1875. That plus what Logerot taught him got NHM into the Southern Arizona Symphony Orchestra when he got back to Tucson.

Still NHM's knowledge of classical music, who composed what, or even what went on in performance was nil. The things others learned in their youth he learned in his forties from the other side of the music stand. He played what they said to play. It was not easy. Fortunately there were five other bassists in the section, four of whom were professionals from the Tucson Symphony section. NHM was the last stand in the section. All the others tried to teach him how to make playing not simply notes but music. Amazing he thought, the way musicians deal with an amateur compared with the way scientists do. Must be something to do with the use of different halves of the brain he thought. Maybe I'll be a more fully balanced person NHM thought if I develop the other half of my brain! In spite of the anxiety that came with each performance NHM really loved being in an orchestra. It was the only group activity he had ever really participated in. It was the kind of effort where the whole was clearly more than the sum of the parts. A special kind of relationship developed between players, each dependent upon the other in ways that were totally unlike anything in his scientific life, and in his entire life until he married. For several years NHM's technical skills grew significantly better and he became much more relaxed in performance. He was even hired to play in a few chamber orchestras, which placed a much greater demand on the refinement of his playing. He felt that he had reached a plateau in his playing, perhaps as far as he could go and he accepted that. NHM had risen in the section from the last stand to about the middle of group, above a number of newer bassists who joined the orchestra. One of them, Ray Luby, had been a professional bassist for his whole career playing popular rather than classical music. He always wanted to play in a major symphony but never found a position that paid well enough to support his family, so when he retired in Tucson he joined SASO. Ray Luby also played an important role in helping to advance NHM's playing skills but not as his other professional music colleagues had done.

At rehearsal one night Ray asked NHM if he had heard the new principal bassist, Patrick Neher, of the Tucson Symphony in his recital. No he hadn't NHM told him, "why do you ask?" Well he said Neher appears to me to be a virtuoso player like the people you have told me you heard when you lived in Paris. You really shouldn't miss hearing him he said. I'd love to NHM told him and asked Ray to let him know when another recital came along. Some months later Ray said "guess what?" Patrick Neher is going to play next week why don't you come along with me to hear him. If I can I will NHM told him. When and where will it be? It's at the UA he said. Neher is also the professor of bass in the music school. And so they went to what was an unexpected delight. Indeed, Patrick Neher was able to play on a par with the most technically proficient players NHM had ever heard. At the time of his performance NHM was the president of SASO and wrote a newsletter for the players called "SASO Notes". He decided to write a brief review of Neher's recital for the next issue. In it he described not only what had been performed but also the superb ability Neher had, and how wonderful the music of a solo bass could be. NHM compared Neher's performance to that of the leading player he had heard in France, Francois Rabbath. Somehow NHM's review came to Neher's attention. He must have been shocked to find someone in Tucson who had ever heard of Rabbath, his idol, and was thrilled that NHM saw a similarity in their playing.

Neher called NHM and asked if he could come to his master class, and bring your bass he said. NHM jumped at the opportunity. Then, once there Neher asked him if he would play something, anything that he knew and liked. NHM had taught himself a few pieces so he performed one of them. Neher fired off a criticism then said he would love to have NHM attend regularly. NHM continued to the end of the term when Neher really caught him off guard. Look he said, I think you ought to play the final exam along with all the others in the bass studio. By doing that you will become a full-fledged member of their group. Forget about how well or poorly you may play he said. Unfair NHM responded, but he did it anyway. Patrick's final success in dragging NHM into his sphere was to request that he actually matriculate in the music school, register for independent study with him and come weekly for lessons not just to play in the Master Class. At that time NHM was the Head of the Cellular and Developmental Biology Department. When he went to registration, his identification card was rejected at the automatic registration desk. The person there said you are not a student and therefore the computer will not accept your registration. Why not NHM said, I have the opportunity as a professor to register for courses. A supervisor settled the issue by requiring that NHM go through the lengthy walk through registration process, which he did. Back in his department head's office NHM got things changed. From then on there were no problems with his automatic registration, which continued term-after-term for seven years. Ray Luby also became a student of Patrick Neher's and he loved it. Unfortunately Ray suffered a fatal heart attack shortly after one of his final exams. Patrick was sick over the possibility that the final exam might have been responsible. NHM tried to convince him that it was very unlikely. Ray loved playing in the bass studio and had never shown any signs of performance anxiety either there or in SASO over the years that NHM knew him. He was a true professional who loved his profession.

The music school administration seemed thrilled with having NHM as a student. Partick Neher turned out to be perhaps the only person in any area who could teach NHM anything. He broke through the shell that protected NHM from the world and forced NHM to do things his way, not NHM's way. He made NHM play at the Arizona Double Bass

Symposia that he organized and to be criticized by the leading bass players and professors of the day: David Walter, Patrick's professor at The Julliard School, Paul Ellison of Rice and USC, even Rabbath himself whom Patrick had gone to study with in Paris. NHM actually got to know the leaders of the band, that is the profession of bassists. He became a member of the International Society of Bassists. In Tucson NHM became known to most of the professional musicians both in the music school and the Tucson Symphony. And he developed connections with a number of organizations that regularly hired him to perform. In various churches he had the chance to play Handel, Mozart, Faure, Bach, and many of the most beautiful pieces he could have ever hoped to perform. He became the principal of the Messiah Sing-in Orchestra, a group that plays each year to an audience of about thirteen hundred who come to sing the choruses as well as listen to the soloists who sing all the gorgeous arias. On March twenty first each year, Bach's birthday, NHM routinely plays in a chamber group for a private party in a foothills mansion in honor of the day. In that house there is a custom built organ that is always part of the orchestra. It was constructed some years after NHM began playing there so NHM had the opportunity to play with it in its first performance. It was the finest organ he ever heard voiced to play with bass strings. The organ builder was there for the premier performance. He came to NHM afterwards and told him how beautiful his bass sounded and that, after organ, bass was his favorite instrument. NHM told him the way he voiced the organ made the bass sound that good. The owner of the organ, himself a scientist as well as a musician, had in his audience people from both of his professions (astronomy and music), as well as from his church. NHM got to know many of them quite well over the years. The same holds for members of various churches where NHM has done a lot of playing. There are perhaps a half a dozen Tucson churches whose members must think that NHM is also a member. Isn't that strange for a Jewish scientist?

The Southern Arizona Symphony Orchestra had only a brief one-year season under the direction of Henry Johnson. The orchestra was formed primarily to provide him with a group to conduct after he retired as the professor of conducting at the UA music school. It was in the middle of that first year that NHM was admitted to SASO. When Henry left Tucson he was replaced by Alan Schultz, the conductor of the Tucson Masterworks Chorale, the music director of a church, a published composer, an organ teacher, and to make his living, a teacher of literature at a private school. During Schultz's fifteen-year tenure as SASO's music director the orchestra became a recognized contributor to the Tucson classical music scene. The music critics from both Tucson's daily newspapers routinely reviewed SASO concerts. SASO produced a major work in conjunction with the Masterworks Chorale each year. NHM served two 3-year terms as the SASO president during Schulz's tenure; and among other things established the "SASO Notes". As you might expect NHM used the Notes to his advantage. One year a rehearsal fell on April first. NHM produced a special issue of the Notes to mark the occasion but told no one of his intention. The lead article was about an unexpected honor that befell the orchestra. In it the claim was made that the Tucson Symphony had been made an offer it could not refuse, the building of an ultra-modern performance hall for it in Ajo. The deal was said to be so lucrative that they had decided to make the move and change their name to the Ajo Consort. Of course Ajo is a tiny place near the Tonoho O'odham Indian reservation where two drums and a flute would saturate the audience. Nevertheless the point of it all was that SASO had been asked to move up so that the large music hall that had been built some years earlier as part of a downtown renewal project would not lie fallow. A beautiful and culturally significant barrio was sacrificed for the construction of the music hall and it could not be allowed to fail. There were many other ridiculous stories as well in the April Fools Notes, but it mattered not. Fully two thirds of the players sat with focused attention on the Ajo story and believed it. Afterwards there was much laughter and someone actually posted copies of the Notes in the Green Room in the Music Hall. The TSO players loved it. Many called NHM to share their response to it. For sometime thereafter NHM always asked his friends in TSO how things were in the Ajo Consort!

After many years of performance NHM discovered to his consternation that the world of the musician is much more parallel to that of the scientist than he had anticipated. Both are dependent upon talent, education, performance, mentors, luck, persistence, dedication, and the ability to ward off the world for the purpose of the profession. Musicians, as are scientists, are tiny minorities in the population wherever they may be. Their careers don't leave much time for other things. It appears to be as hard for a young musician to get a job in a symphony orchestra as it is for a young scientist to get one in a college or university. Money is always a problem to find or to earn. Marketing yourself appears to be difficult in both professions. In some ways talent traps both kinds of people. Both life forms involve creative endeavors. Yet once into the work aspect of the profession symphony musicians simply play whatever is put on their stands just as scientist/professors teach whichever courses are assigned to them (the content of which is either dictated by competitive national exams or by prerequisites for higher level courses.) There appears to be an underlying artistic factor that drives a similar kind of person into music and science and perhaps to other pursuits such as fine art or writing as well. People of this type are motivated by factors that differ from those of the majority. There is some kind of force driving these people that makes it hard for them to give up what they do even in the face of failure or utter necessity. Creativity and appreciation of the product that they themselves produce are what keeps these kinds of people happy. In this regard, NHM's music colleagues are the same kind of people as his scientific colleagues. He is sure that most practitioners of either music or science never dreamed of their similarity to one another. NHM's bass teacher Patrick Neher is the exception. He once was quoted in an article entitled, "Mendelson mingles music and molecules" that appeared in the Arizona Daily Wildcat on March seventh 1994 as saying, "He's one of my best students. Science and Music are similar disciplines. A research scientist and a musician are both artists." How would he have known that? Simple, both of his parents were scientists as well as musicians. No wonder he can do just about anything.

The void in music that accompanied NHM to college has finally been filled, at least to the level where he can contribute to a beautiful product that he enjoys as much as the audience does. In some of the ensembles in which he performs, he has the best seat in the house even though he stands to play in order to keep his hip from total destruction. Playing the bass destroys the body, as playing any instrument does. The body wears out with time as a result of repetitive stress injuries. Conservatories know this and now teach ways to prolong the playing life of their graduates. NHM has made many adjustments based upon their new understanding of the problems. One of the dangers however reveals a lot about how the body and the mind work: during performance all pain is gone. And for quite a while afterwards you feel absolutely great but tired at the same time. When pain returns there is an inclination to play some more to get rid of it. The narcotic of performance is much like other narcotics. It permits you to do more damage to yourself without feeling anything except wonderful. There is of course a breaking point but fortunately NHM has not yet reached it. He has seen others who have. It is not pretty. Both knees replaced in a cellist. A neck bone replaced by a transplant in a viola/brass player. Joints, nerves, and muscles all take a beating. Deafness reigns. Some say everything has its price and question what's left of a performance after the sound fades as if you have wasted your body for nothing. True you give something away of your body but what's left is more valuable than a physical object. There has been enrichment in your life and others by what you have done and that remains even without your having to recall the performance. It's a kind of a spiritual high ground that permits you to go on in life at a higher level. It's not something you need to tell anyone about. It's just there. And it sure beats the void.

NHM is sure that there are easier ways to go through life than the path he took even given the missing things of his formative years. He does not see his route as one of the perfect mistakes of his life, at least in terms of its logic or execution. Yes, he started from zero but he went to max in understanding modern biology. Yes, he erected a shell to block all outside influence but within it he created his own world and experimental system. Yes, he developed a plan by age eleven to escape and it worked well enough for that purpose. No it wasn't the total solution but it provided a bridge to it. Yes, he has been very angry at both his own ignorance and the achievements of others with lesser talents, and he resented never sharing in the riches they received. The anger wasn't the mistake. Not valuing what he did get is perhaps closer to it. Yes, he survived two years in the military, but he learned so much about the character of America then that it was well worth the risks. If you had any idea of what went on in the Army it would make your hair curl and Abu Ghraib look like kindergarten. It's not an honorable thing to be proud of. Yes, NHM fought against many things along the way not necessarily to defeat them but rather to give himself some space to live and to pursue his goals knowing full well it would have to be in coexistence with theirs. On and on he went without really seeing the light. And because of that he now pays the price of his perfect mistakes.

Here are the ten Perfect Mistakes that constrained NHM's existence as forcefully as any ten dominant genetic mutations could have:

1. NHM mistook the basic intention of people to be at least neutral if not positive, and therefore expected a fair chance in his pursuits.

2. NHM assumed that one's parents or guardians would help their offspring to understand the world and how to survive in it.

3. NHM thought that if you could do something you would have a chance to and if you were better at it than others that would work to your advantage.

4. NHM believed that you should do what you liked to do rather than what you had to do to achieve a particular goal such as bringing greater rewards to yourself.

5. It never occurred to NHM that having a lot of money was vital to protecting his own interests.

6. NHM refused to abandon honor in order to defeat his enemies regardless of how dishonorable they might have been.

7. NHM was certain that it was possible to work alone and survive.

8. NHM looked upon mentors as a sign of weakness.

9. NHM was willing to learn only that which interested him.

10. NHM never realized how much he shortchanged his family until it was too late.

13. The benefits and the costs of perfect mistakes.

How can there ever be any benefit from a mistake, be it a perfect or an ordinary one? Only by chance perhaps, as for example when a blind man takes the wrong turn but avoids disaster by having made the error. Clearly a mistake in one context isn't necessarily a mistake in another context. In the case of the blind man, survival was obviously a more significant benefit than the inconvenience of going the wrong way. One might view NHM's path through life as comparable to that of a blind man by equating the void in NHM's life with the lack of vision in the blind man. In both cases chance plays a role in the outcome of every decision made to a much greater degree than it might for someone with neither of the deficiencies. Like the blind man NHM too survived and luck must get a fair amount of credit for it. Every choice that he made could easily have led to disaster. NHM sees that now although by rewinding the film he also sees that at the time little thought was given to the possibility of failure. That part is quite frightening to realize given that it wasn't just NHM's survival that was at stake. Where was the responsibility for those he was responsible for? Did NHM in fact perpetuate his own perfect mistakes? Was not his own escape the creation of a world from which his family then could not escape? It is possible that any benefits that NHM can list are results of the most selfish things a person could have done, a cost far in excess of the benefit. What if anything can be said to be a benefit from what NHM has done?

Not much. Wouldn't perfect mistake number one, the idea that people are basically friendly and helpful, become only a simple mistaken assumption once you realize that survival is everyone's goal, fairness doesn't come into the picture? Yes it would, but in NHM's case it didn't. He wasn't able to give up the hope that basic intentions were positive. He held on to this idea by assuming that it was just bad luck that it hadn't proved to be so in the vast majority of the people he had to deal with. By keeping the ideal alive in the face of reality NHM adopted caution as a basic principle. Never assume you will get a fair chance became his expectation. And he believes that served him well as a kind of preemptive strike against the norm of rejection; the fact that people put him down in order to gain the advantage for themselves. Perfect mistake number one proved to be of adaptive significance although it could have easily been lethal. The cost of this perfect mistake came when trust was placed where it shouldn't have been. Plenty of examples have been given in earlier sections. They are but a small subset of the penalty that NHM paid at every turn of the road. From his reading of history it appears to be a price that has always been paid by those who trusted in something: a price of the human condition.

NHM's fallacious assumptions about what older generations pass to younger ones, particularly in families seems to him must be true for most of those living in enlightened civilizations that have, as in his case, failed as a result of war. War's disruption is felt in many ways both near and far from the battlefield. One sees in his case the ramifications of removing the most capable member of the family (even including all of his flaws) and leaving the rest to fend for themselves with minimal resources, either economic or cultural. The young have got to make do on their own on the street or in the jungle, no real difference. Sure environments differ, some better than others for survival or acculturation but it really is hit or miss where chance and necessity play out from moment to moment. Those for whom the assumption of perfect mistake two, that survival knowledge passes in family lineages is true, are jump-started in the world. All others must trail behind them regardless of their innate abilities. For those who make the mistake as NHM did the only benefit is that if you live, you know you can. Afterwards having very little is not perhaps so frightening as it might be. What will become of you may not be clear but the confidence that it will be something becomes very strong. Knowing that it's not the best way to get started is surely the major benefit because it carries over to the next generation where you appreciate what the difference can mean and work hard to convey whatever it is that you know and think can be helpful to them. The writing of a book like this one is a kind of benefit from perfect mistake number two.

The idea that ability alone provides opportunity and reward probably leads to more unnecessary disappointment and frustration than it should.

Everybody has abilities at some level but not necessarily the ability to properly assess where their own ability ranks in comparison to someone else's. Even if they could, ego throws a terrible monkey wrench into the equation. Who hasn't confronted someone else's performance and thought I could do a hell of a lot better than that? Maybe so, but that is irrelevant. What matters is whether you have the opportunity to show that you can. There appears to be just so much space in the world for every kind of talent. Whether you occupy it or not is not governed by the talent itself. Not by a long shot. If you can't or refuse to recognize this as in NHM's perfect mistake three, you just keep plugging away as if your lucky number will eventually come up. If it does you may be in the spotlight for just a moment, that's the benefit of perfect mistake three, but is it worth the cost? It hardly seems as if it could be. Possibly you get just one moment of glory but you definitely get a lifetime of being passed over by the spotlight that comes to shine on others. When the light is in your face it can blind you in many ways. To NHM the most dangerous way is to create the false security that you won because of your ability and thus forget that those who wish you to be in the dark and they in the light are not deterred by your momentary illumination. "Watch your back" is what they used to say, "ridiculous", is what NHM thought. That was his pure stupidity. An awful lot of things could have gone much better for him had it not been for this arrogance.

Doing what you like rather than what you have to do to achieve something ought to be the most perfect recipe for failure known to man. To believe otherwise cannot be anything other than a perfect mistake, as the case of NHM's perfect mistake four. Failed indulgences are probably on a par with failed marriages in terms of their frequency. How can there be any benefit from acting out this self-destructive desire? The chance that luck will provide the rescue is so unlikely that NHM won't count it. That leaves only one thing: someone else will have to do the labor for you. That is the burden that fell upon NHM's wife, and later also his daughters. It is the biggest debt that he incurred in his life. The fact that he didn't even recognize the debt until it was too late to repay it ranks as his most terrible perfect mistake, number ten. Yes, one perfect mistake can lead to another creating a network that can never be corrected! But that doesn't mean that one can never recognize the mistake. NHM knows how wrong it was to incur this debt. Whatever he achieved was at the expense of others. How could anyone without independent wealth have ever gotten the idea in the first place that you do what you like, period? It hardly seems possible to hold this view given that survival was the core of NHM's existence. Could it be that he got it backwards, that really he just happened to like the few things he could do and mistook that to mean you just should do what you like? Could he have mistakenly interpreted the fact that no one ever suggested what he should become when he was young and it wasn't quite clear if he could indeed do anything to mean that he should just do whatever he could? Was it possible that just surviving the war years meant that you were free to do anything? What rings the loudest in NHM's mind is a comment made to him one day at a social gathering of lawyers who were colleagues of his wife's during the years that she was an Assistant Attorney General in Arizona. One of them asked NHM what he really did as a scientist and professor, and he said that he played with mud pies in his laboratory forever and someone else thought what he found was interesting so the government kept giving him money to keep playing. "Nice work if you can get it" the lawyer replied, and all the others listening to the conversation nodded in agreement. Now NHM has got the dirt on his hands.

How could anyone without it miss the need for money? NHM did. That was his fifth perfect mistake. Working from the eighth grade through high school he earned enough money to afford buying his first bass and also later his wife's engagement ring. In college with his father's military disability paying his tuition he was able to cover the rest by working in his band. But when he got married his mother-in-law asked, how can you support yourself and my daughter? We don't need money, he told her. How will you be able to buy shoes and the like she asked. We'll have enough he said, if we don't I won't wear shoes. They managed in graduate school. And NHM did inherit one hundred dollars when his maternal grandfather died, the only thing he ever inherited! Wow, they bought a turntable, an amplifier and a speaker for that. Later they got some records and could play them. Money seemed adequate then, and in the Army that followed, and in London with NSF support and after that while a professor in Maryland too. The curtain came down however when they got to Arizona. Yes then NHM knew they really couldn't make do with what he could earn. He took a course and got a license to sell real estate hoping to get a second job but his wife thought he would be wasting his time and talents. So she went to school, became a lawyer and kept them financially sound. That was her survival move for them. But it wasn't strategic planning so to speak and it too was too late. Now it is clear to both of them that you need much more than they have in terms of assets to be able to afford a decent end game plan and to provide backup for your children and grandchildren. There's no way to get that now. There can be no more perfect a mistake than number five, failing to realize that you need to acquire a lot of money during your working life. There is absolutely no beneficial outcome possible from perfect mistake five.

On my honor I will do my best to...." starts the famous Boy Scout's Oath. It stuck in NHM's mind along with a whole lot of other things learned there that substituted for things not learned elsewhere. When a grease fire engulfed the stove in his mother's kitchen while she was trying to cook something and she used the fire extinguisher on the wall to try to put it out, NHM realized that carbon tetrachloride generates poisonous phosgene gas when it touches hot metal. He quickly unlatched the kitchen window, got everyone out and called for help. On the first fire truck to arrive was the man from whom he had taken firemanship merit badge. NHM told him what happened. He attached a smoke thrower to the open window and evacuated the air inside the kitchen before anyone was poisoned, then went in to extinguish the fire still burning in the stove vents. That was when NHM was thirteen. From his Boy Scout era honor became internalized, as did phosgene. With honor he felt as if he commanded the moral high ground. It seemed to have worked. How then does honorability become a perfect mistake? NHM identifies it that way because it creates a class of abuse you must suffer without due recourse to fight back. The military principle of complementarity (use A against an A) must be abandoned forcing a disadvantage upon you in the conflict. Carried to the extreme, honorability as NHM defined it in perfect mistake six can be lethal. That is fortunately a cost that NHM has not had to pay. Having escaped it, the benefit is clear. You walk away from your enemies knowing you have attained a higher degree of civility than they have. There is strength in honor and you know it the next time another conflict arises. Never forget however that the ruthlessness of your opponent can defeat you in this game. When that happens you'll know why dogged honorability is a perfect mistake.

There were many experiences in NHM's life that firmly implanted the idea that not only could you work alone and survive but that doing so was desirable. When there is no team; teamwork isn't an option, so you go with what you've got: yourself. Never mind that two heads are better than one, or that six men can lift a bigger stone than any single one can by himself. NHM didn't care about such things. What someone else could do didn't interest him. What he could do was all that mattered. NHM struggled to invent everything he needed and wished many times that his intelligence was ten points higher than it is. That would have saved him a lot of work. He could have seen relationships much more quickly and gotten much further than he did in the time given by his biological clock. Instead NHM guarded his turf, fighting to keep it inside his protective shell as long as he could. The shell thinned with age however and was eventually penetrated more than once. It didn't take long for those insiders to discover that NHM didn't know anything but what he knew from within. NHM was sure they had never confronted anything quite like it, yet they adjusted to it all the while trying to enrich it. It was generous of them and NHM benefited tremendously from it but it could not shake his innate need to get away from as much of it as he could back into the inner shell. As if in the center of a Zuni figure, he had to wander around with ideas in his shell until he digested them in there not just with the final idea that you are going to die as the Zuni do in theirs. That might be a true benefit of perfect mistake number seven; alone the brain can function with limited noise to distract it. To a neutral observer wishing to maximize output it would be a trivial benefit compared to the cost of not having all the synergistic interactions that come from working together to say nothing of the social benefits gained as well. Civilization is not working alone and neither is survival in the real world.

To this day NHM is not sure that there are any truly successful scientists who weren't helped in some way by mentors. Whether that is so depends of course upon the definition of success. NHM would judge success to be having made discovery, being recognized for it, and sharing in the rewards for achievement that others of like accomplishment receive. NHM is not talking about being the most successful of the successful but just being somewhere in the middle of the pack of equals. What experience has taught him is that there are a disproportionate number of those at the top of the pack whose accomplishments and or intelligence really don't warrant the success they enjoy. What these people excel in appears to be the skill of finding and using mentors. Applying the principle that people almost never do anything that doesn't in one way or another work to their own advantage NHM has often wondered what the mentors get in return for their services. The flip side of this is that the mentored must incur a debt for services rendered. In NHM's own particular case using mentors was never really an option. For one thing he couldn't think of any currency he had that could be used to repay the debt save the discoveries that he made. To give that away really would have left him with nothing at all. Who knows if you can ever make another discovery? In view of the fact that no one ever stepped up to volunteer as a mentor for his scientific career he concluded that to get one you would have to ask for the help. You would have to present yourself as someone in need to someone else who by definition would have to be your superior. No, no that was not for NHM. A scientist faced with death as the geneticists were in the Lysenko era in Russia could be excused for giving up his discoveries, his "Scientific Being". Short of that NHM couldn't see swallowing the bitter pill of either asking for help or giving away the fruits of his efforts in payment for getting it. NHM's eighth perfect mistake curtailed the sphere of influence he could ever command in science, even in the small corner of science that he carved out for himself. Put simply that meant that neither his ego nor his pocket book could ever rise to satiation. The ego part was easy to swallow because any praise given to his productivity at all was genuine. The stain of the economic part is not so easily washed away. There is no way to compensate for it and thus it compounds.

In a world of misinformation and disinformation you wonder how being disinterested in anything could be a perfect mistake. NHM can answer that. Think of what he told Paul, the student from Massachusetts who wrote to him in 1992 to find out about discovery and invention. NHM counseled him to try to learn as much as he could about everything in any way that he could so that later he would be able to assess what he knew in relationship to what others already knew. NHM said that to him not so much to discourage him from just rediscovering something but to protect him from the humiliation of being told afterwards that is what he did. There is a time when being focused can be of great value but operating in a true vacuum right from the start can't be the ideal case. When NHM was Paul's age he had a lot of questions that he wished could be answered but he just didn't live in an information rich environment. Quite the contrary, he would describe it as a mix of the blind leading the blind and the opinionated representing ideas as truth. There are just so many false starts a person can endure before coming to grips with reality: best not to heed the advice others give. From that comes shutting out all inputs save those of direct concern to your interests and needs. Once in place calmness emerges as the babble subsides, a calmness that helped greatly when it came time later to fill the void. There was power in being able to pick and choose what it would be that you would take in. It was like a triage of information. This is something I want to know now, this is something I might learn later given the luxury of time, or this is something I probably will never want or need to know, case closed. So NHM picked and chose to build the person he became, deep in a few things but very shallow in all the rest. So what? How could that be perfect mistake number nine? NHM sees it that way because the edifice built was limited by the range of things known when the initial decisions were made concerning which information belonged in which category. He had no way of knowing whether those choices resulted in what could be called the highest and best use of the land, in this case land being his own innate abilities. Would NHM have been much better suited to something else in life? Doesn't everyone eventually come to ask that question? Experience has taught NHM that there are many other things that might have interested him as a thing to do in life but by not knowing that they even existed when his self-construction was in progress they never went through the triage process. Having been missed they couldn't possibly have served as foundations upon which anything could be built later. That is what makes mistake nine, learning only what interested him, a perfect mistake.

Perfect mistake ten NHM guesses is the kind of thing that ought to be taken to the grave without ever being said, much less written down. It's the kind of worst result that can come from misguided intentions. Exigencies notwithstanding, course corrections should have been made to rectify the wrong and they weren't. As if NHM were trapped in quicksand his inertia resulted in the grand badge of dishonor: forcing all the others to live his deprivation. Not surprisingly, they fought their way out in their own ways, with no help from him. How much they might have regained cannot be measured, as a scientist would wish, particularly one responsible for it hoping to see it minimized. This fact justifies its inclusion in the "perfect" category. One's escape becomes another's trap as if the physics of space constrains existence along the genetic lines of descent. The two great sources of information in humanity, genetic and cultural, entwined forever in each lineage. NHM thought he broke the link in his own case, but did not foresee that he would impose his own upon the next. No, he didn't break any links he just changed their twist. What hurts the most is not having applied the principle that less may be more. NHM should have done less for himself and more for them. Too late now.

There remains one last thing to interpret, where did the humor come from and what was its purpose? It definitely wasn't part of NHM in Brooklyn, or Park Chester, or even when living near LaGuardia airport, ages four through twelve. Nothing was funny then. Survival was still an uncertain issue and all other efforts were focused on how to achieve Plan A at that time. Once during that period NHM's cousin Chuck came to visit for a day with his parents. NHM took him out on the street. In their wanderings about they passed the house of a boy NHM knew. His older brother was a laborer who had an eighteen-foot bull whip, which he had taught NHM how to snap. NHM asked Chuck if he had ever seen one. Oh yeah he said. Do you know how to snap one he asked? Oh yeah Chuck said again. Hang on I'll see if I can borrow it, NHM told him as he went to ring the bell at the apartment where they lived. The kid he knew came to the door. His older brother was stretched out on a couch trying to sleep something off. Can Neil borrow the whip he asked him? Yeah, but tell him to be careful with it and to bring it back before dinner, he responded. So out they went with it. NHM snapped it a couple of times to show Chuck how he could knock a can off a fence post near the apartment complex then passed it to him. You know how, right, NHM asked. Yeah, Chuck said as he laid it out on the street in preparation for a snap. He actually looked professional as he got set up. Then he gave it all he had and the whip took the seat out of his pants and left a red welt across his rear. Oh no NHM said, are you alright? It burns awful, Chuck answered. Observer remembers this incident because he thought NHM would burst into laughter but didn't. It is the first time Observer can recall NHM thinking something was funny. The reason NHM didn't laugh was because he and his cousin both knew that Chuck was going to take hell for ruining his new pants. They quickly returned the whip and limped home. NHM remembers Chuck going to face the music like a man. He wondered what he would have done had it been him. In a strange way his injury saved Chuck from probably much worse punishment had it been only the seat of his pants that were ruined. No, humor had not yet become part of NHM.

When NHM became big enough in high school to feel that he could constrain his father's explosive and sometimes violent temper episodes, he appears to have started using humor as part of his armament. The first time he came close to laughing in his father's face was a step in the development of his humor tool. NHM's father had insisted that he work with him on the weekends to paint and exchange winter storm windows with summer screens on the sixty six windows in their large house. They were working in his sister's bedroom where two windows opened onto the roof of a porch like room below that was part his father's home office. NHM's father was out on the rooftop painting the outside of the frames and NHM was working inside doing the same job there. One little strip on the inside was left unpainted so that when his father climbed back into the room he wouldn't smear the paint and get it all over himself. The time came when his father had to climb back in. He did so facing backwards one foot in place on the outside roof the other coming through the opening and as everything he did, moving slowly, cautiously downward feeling for the floor beneath him before transferring his weight to it. Trouble was his foot was headed directly for the can of open black paint just below the window. NHM said not a word, knowing that his father was well aware of exactly where the paint can must have been and believing that he would at the last few inches make an adjustment to his angle of descent so as to avoid disaster. Slowly but surely his father's leg made its way in space as if his foot could see. Then, unexpectedly, instead of moving away from the can in the same cautious way, he all at once slammed his foot down into the can, felt his mistake and kicked the can off as if it were the jaws of a monster trying to devour him. At least half a gallon of black went over onto the floor by the time he got the rest of his body in. He was horrified; NHM was nearly crippled with trying to prevent laughter. His father lunged to the floor and with his arms created a dike meant to contain the expanding circle of paint. At the top of his lungs he grunted: "rags, rags, rags". NHM's mother on the floor below at the time heard only a grunting sound and shouted up, Michael, are you calling? Rags, rags, rags was all he could say, so NHM added better get some rags up here quickly there has been a paint spill. She came up but had no rags to meet the need so she improvised by throwing some bed sheets to him. He rejected them shouting rags, not sheets, rags, rags. Use the sheets Michael she said, I don't have any rags. Trying to do so he lost control of the dike allowing the paint circle to expand rapidly to the size of NHM's sister's bed. His father would allow no help so NHM ran out of the house and laughed until he cried. The invincible man was not perfect. The black stain remained in spite of all efforts until they sold the house on the way to his new job at the mental hospital in New Jersey. Not only that, NHM lived through the incident and learned that some things really made him laugh. It was the beginning of a new trait.

Like any new tool, NHM played with his sense of humor as much as he could, and some of it was indeed in very bad taste. He won't go into the details but suffice it to say he tormented his father on more than one occasion, once even proving to him, a psychiatrist, the power of the mind over reality. NHM didn't realize at the time that much of what he did was to tempt fate with humor. He began to do things that could lead in directions that were unknown, perhaps dangerous, and every time he was able to control the situation afterwards, he became more and more confident that he could survive danger. This went on for years but by the time he went to college he had discovered the most important use of humor in his life: the fact that he could use it to reinforce his protective shell. With it, NHM could keep people at bay, uncertain of what he might do or say next. He developed the skill of getting people totally off guard by using humor, or defusing anger with humor, and even covering his true beliefs and intentions using humor. Many of those who loved the humor never realized what he thought about them or their beliefs. Humor was a way NHM could discover a great deal about people. The way they responded, what they had to say, and the like gave him a great deal of information about people that they never knew they were revealing. As he honed these skills he enjoyed it more and more. Humor was a way to stifle his own frustrations, to establish false impressions, to divert attention or to focus it, to seize the upper hand. With it NHM could shut down others when their babble became overwhelming to listen to. It could in an instant take the wind out of the biggest windbag and set the record straight; the smarter the opponent the bigger the prize. Just like in fishing. In a somewhat more refined way NHM could use humor to let people know just how smart he was, or how he viewed the important things of life. This form of NHM's humor can be found in many poems that he has written to mark events in people's lives. If NHM lives long enough he will try to find some good examples and put them into another book.

14. Keep your eye on the clock.

There is a time and a place for everything the famous adage says. Believe it and act accordingly. Aspirations are one thing but achieving them is another, and that's where some appreciation of time becomes important. How far ahead to look is of course the issue. The more problematic the environment, the shorter the distance into the future it is possible to look. When things aren't great, maybe you take it day by day. That works. But in NHM's case the unfortunate thing is that once that truncated horizon became established it became impossible to readjust it later to something more appropriate to the new times. And so it went for most of his years with never a three, five, or ten year plan, never benchmarks to be reached by time x or y, no long range plans ever. Then one day it came time to plan for retirement, to plan for an end game plan, to play catch up on the things dreamed of but never attempted. NHM knew this too was something he did not know how to do, so he decided the best thing to do was to give himself as much time as possible assuming there were many mistakes that would be made and from them perhaps eventually he could get it right. Step one was to search for a summer home. A place to escape the desert's summer heat and to which children and grandchildren could come for some family time, was the main idea. Perhaps even a place where he and Joan could do other things, as consumers for a change. A wet place rather than the usual dry one seemed like a good idea. A safe place, as well. The specification list grew until it appeared that even heaven might not do. Joan's sense of realism took over. We must go and look on the ground to find the suitable thing. We have studied it long enough. And thus began a ten year series of summer visits all meant to get them somewhere from here. In parallel, planning was started to examine how their retirement finances might be arranged and how they would know when they had sufficient funds to stop work. In this NHM's role was somewhat marginal having relinquished control of all financial matters to his wife when she became a lawyer. Like it or not the job had to be done and they did it to the best of their abilities.

Here they are in Tucson Arizona for 38 years but in truth they probably know more about London, Paris, Cambridge, and Lausanne than they do about Tucson. Or about Arizona, or the West, or just about anywhere save the places where they grew up, went to school, or spent time in the Army. NHM felt about as shaky knowing where to begin as he did when he had to choose a college. If we want water how about the Pacific Ocean they reasoned. There's logic. Off they went to see what it was like. Southern California was unaffordable, congested and unattractive to them. Cross it off. Seattle was a concrete jungle, dismal and gray, not exactly "On Golden Pond" territory. Cross it off, and so too the rest of Washington's coast. Twenty-four hours in Vancouver was a terrible joke, except for the fact that they found a badly needed living room couch in a furniture store they passed by chance. It is the most comfortable piece of seating they have ever had. The joke came in the private club they stayed in. As an esteemed member of the famous Cosmos Club, NHM has the ability to stay at private clubs around the world and booked them for one night at this place. The drive from Seattle was a little longer than they anticipated so they arrived a bit late. "No problem" the club had no guests but them. It had just been renovated and they were the first to use it. Great. It was a big beautiful room with a luxurious bath. Small problem brushing their teeth however, they could get no cold water. Now in Italy and elsewhere they had stayed in many places with no hot water but this was a first. NHM called the desk and the man said "no problem". We will move you to another room. They got dressed and changed rooms. Undressed again and, no cold water again. NHM started to investigate. The water in the toilet was steaming hot. Not a good sign. NHM called the desk to let them have the bad news. Where is your hot water boiler NHM asked? I can't take you to it, he said. "Was it part of the renovation?" NHM asked. Yes it was the clerk informed him, they had rushed to get it all hooked up earlier that afternoon. Look, NHM told him they have made a mistake. Perhaps I can correct it at least temporarily. No sir, we can't do that. OK, What can you do? Well we will move you into another private club next door. So they got dressed once more and dragged their bags to the neighboring club, a shabby place that had cold water and some hot water. They flew that coop as soon as they could the next morning, found their couch and headed back to Seattle. Problem was that at the border they wouldn't believe that they had been only twenty four hours in Canada, and even if it were true what could they have done there. The obnoxious INS people delayed them for several hours. By the time they got to Seattle their reservations were all screwed up and there were no rooms to be found anywhere. The Queen of England was there for something or other and all rooms were booked. At 2:00 a.m. NHM finally made contact with a very fancy private club, also with reciprocal arrangements with the Cosmos Club and they rescued NHM. The rooms in this club were suites as large as their entire house in Tucson.

Next stop, the coast of northern California and Oregon. Both were beautiful but the costs in Oregon were much less than in California. On the way there they stopped for lunch in a small town, Ashland. There by accident they found the famous Oregon Shakespeare Festival and managed to get seats at the last minute for the production of "Coriolanus". It was spectacular and NHM learned a lot about himself from it. By comparison the coast looked drab. Yet they found a nice place and were within an hour of purchasing it when they decided to take one last look at the oceanfront. The wind was fierce. They were sandblasted so quickly that they both realized that their grandchildren could never enjoy the beach at that location. Back to Tucson via Jackson Hole in case there might be something of interest there. There wasn't. The West Coast didn't work for them. The question was if not there where is a place that will work? How about Montreal? After all it's got both British and French culture. Yes but. The two halves hate each other. How about north of Montreal in the Laurentian Mountains? Sorry, it's pathetic. Back to Tucson they went. Hey, they were told, what about the beautiful Apostle Islands in Lake Superior? The town of Bayfield, WI might be right. Indeed it was beautiful and they made a deposit on a new cabin being built on its shore. Back to Tucson to make plans. Over the winter NHM followed events in Bayfield. The weather was abominable. The snow too deep to allow fire trucks to save houses that burned down. The duration of the season for them would be very short. Who would take care of their place the rest of the year? If something went wrong how could they fly in? By car it was a three to four day trip. No it wasn't the place for them. Fortunately their deposit was returned to them. Their plans weren't going too well. What about going back to Europe? Wouldn't that get to be a drag and an expense? Perhaps we are too old for that they thought. If not now they might be soon. They could have lived nicely in London, or Paris, or Cambridge, or.... but what about their children? It wouldn't be cheap for them to visit. Forget about Europe even though we belong there they concluded. Living in Europe is just history for us now. We are back in Tucson. We are nowhere.

Maybe we should go back to Ashland, Oregon they thought. It did have something to offer. OK let's visit it again. Theater was as good as ever but there was nothing they could afford there. Like California it was out of their range. What the realtors said was come up and rent for two or three years until you find the place you want. When it comes on the market, if you can afford it, buy it that day. They couldn't do that. Neither of them was retired yet so Ashland was out. Before they returned to Tucson they visited a nearby town, Jacksonville, on the off chance that it might be a place they could consider. What a lovely place but probably too costly also they thought as they drove around. Look there's a local real estate office; let's look in the window, NHM said. It was closed but had things shown that looked like beautiful country homes. The door popped open and a woman asked if she could help them. She asked them to come in even though it was closed. They told her what their situation was and she told them what it was like in the area of Jacksonville. Then she asked, would you consider a small farm in the Applegate? A farm? They looked at each other. Plan A reappeared fifty one years later. I live right near it she said. If you like you can follow me to it to take a look. Neither of them could believe their eyes. It was perfect. They made an offer the next morning. In 1999 they bought a farm in Oregon, forty five minutes drive from the Oregon Shakespeare Festival. A farm after all, NHM thought. And so they start again in a new life. They are farmers in the dirt, not gentlemen farmers. Everyone in the Applegate Valley thinks they are insane. They have no mechanized tools. They work by hand. Their neighbors knew that they did not know what they were doing. NHM and Joan also knew they did not know what they were doing, but they were learning. Bit by bit they converted the meadows into lavender growing fields. They converted the Red Barn into a lavender-drying barn. They converted the large garage/workshop into a multi-purpose building housing a music studio/guest quarter with kitchen and bath, and a workshop/garage that could be used to store one of their cars in over the winter when they were in Tucson. They rebuilt much of the old farmhouse with furnishings from around the world. They closed up the old gold mine located under the meadows for the past seventy five years. They had a new forest path built that leads from the farmhouse to the creek at the edge of their property. They repaired fences. They planted new varieties of lavender each year. They harvested by hand, made sachets of dried flowers, had oil distilled from fresh flowers, sold dried flower bundles. They created

Applegate Valley Lavender, LLC. The quality of their product became known in the local region. It was hard to keep up with the demand of customers. They loved it, but it was very difficult to do for people of their age. They did not keep an eye on the clock and it caught up with them.

They proved they could do it, but they also proved that they couldn't do it. Acting like children rather than grandparents can only go on for so long. The weak links start to go and all of a sudden they had to give up the farm. They had no choice. No one in the family could take it over. Having others do all the work would be way too expensive for them. Fifty one years was too long to wait for the start of Plan A. They should have known that but the beauty of the place clouded their reason. In the summer of 2005 they found a person to buy their farm. For the buyers it was a dream come true; a lucky find of something just getting going on someone else's labor. Fleshing out Plan A worked although it hadn't yet reached the point in NHM and Joan's hands that could be called a success as an agro-business. Nobody could yet live on what they could produce but it could have developed into that. They both knew from the start that they couldn't last forever on the farm but they hoped they could stretch it out for ten years. Well, they got about six. Not bad. Those who watched them build it are quite sad that they had to move on. They are sad too but ready for the next adventure. In a way they retreat now into the inner world of their minds hoping to find a realistic end game plan that they can enjoy. Neither of them is hell-bent however on just getting fat. A sequel to this story is not yet therefore totally out of the question. There is no crying over spilt milk either. They gave it their best shot and have gotten many rewards from it. NHM has proven that Plan A conceived at age eleven was a viable plan for him. It would have been a lifestyle very well suited to his needs. The default Plan B worked too but having now tasted Plan A NHM is left with the impression that A might have been just a little bit better. NHM guesses he wasn't that dumb at eleven.

15. Guardian of the core reveals his secret and speaks his mind

You have heard the story now from the voice of Observer and have seen just how clever Creator is but Guardian has remained quiet in the background surfacing but once or twice to comment. Guardian now asserts that he must in good faith say no no no no no no and no. And Guardian wishes to reveal what he has actually done not what his partners' think his job has been. Guardian was never just a place where currency or valued possessions were deposited to be locked away out of sight, out of mind and safe, he tells us. Nor was Guardian a passive partner in what he accepted and kept forever, not at all. I actually shredded quite a lot of things over time, unlike Observer who had to retain everything always. So the core itself was my business not just the keeping of it. I was partner to its building as well as the editor of what it came to contain. As such I feel it my right to defend its contents. I strongly believe that there are no "ten perfect mistakes" in NHM's core. Had there been NHM could never have had such a rich existence. What Creator has identified are not defects in the core, they are defects in the world. What influenced the course of events in NHM's life did not come from within it came from the outside. Had Creator been able to look into the core he would have seen how finely crafted it is. It is a robust creature with plenty of redundancy and fail safe mechanisms. That is what permits it to survive the onslaught of insults the world delivers to it, and that is the reason NHM could achieve what he did.

I have no argument with the facts as presented by Observer or the logic of Creator's analysis, Guardian tells us. Their error lies in the way they view the overall equation. Theirs is a one sided view, pointing, in my opinion, in the wrong direction. It isn't sufficient, as they have done, to conclude that the only thing you need to know about a product are the reactants that produced it. There is more to it than that. There is the environment in which the reaction goes on. Had they realized that they wouldn't have jumped to the conclusion that the fault could only be within NHM. Had they, as I have done, examined how NHM's core would play out in a different environment than the one he faced, they would have realized where the defect was. Had they, as I have done, examined how NHM's core would play out if he had lived at a different time, they might have realized where the defect was. It is disingenuous of them to point a finger at me and place blame where it does not belong. Read what you have said are the ten perfect mistakes again, I say to them, and ask yourself whether they might not have really been virtues had circumstances been just a little different than they were. That said, I accept full responsibility for Perfect Mistake ten. I should have recognized and rectified this terrible mistake in a timely manner but I didn't. And so the core that I guarded was significantly flawed. How far into the future this flaw is retained in NHM's lineage remains to be seen.

Thank you for hearing all of this. But is there a take-home lesson? NHM thinks so. Neither physics, nor genetics, nor even molecular biology can yet provide you with a likely scenario of what your life will be like. That being the case you have no choice but to go with what you have been given, your genes and your culture. Mistakes are inevitable, sometimes even perfect mistakes. Why shouldn't they occur? Even genes suffer these indignities. It's part of life at a very fundamental level. Accept those that come your way and hope for the best.

LIFE MAP

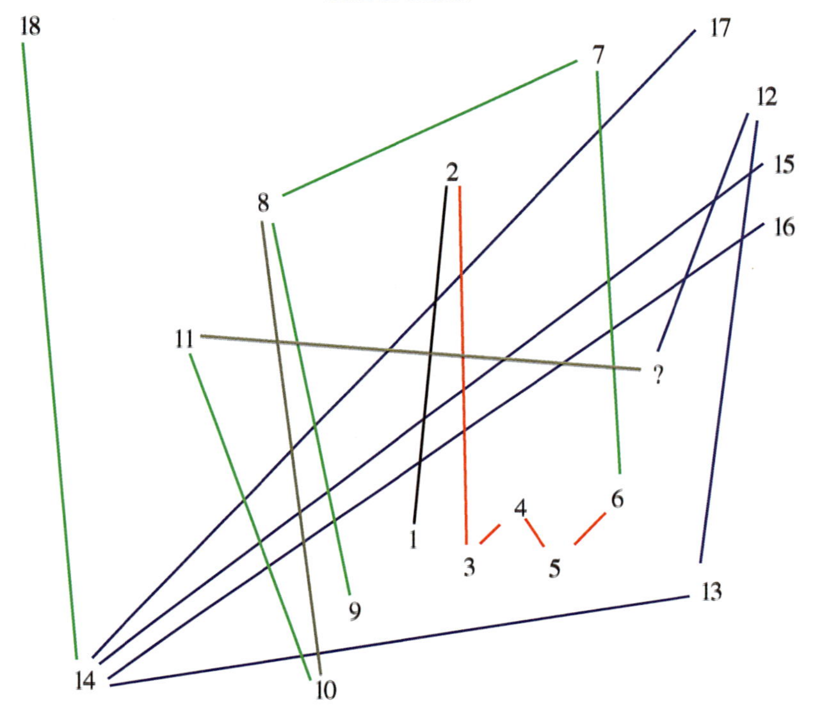

1. 1937 Manhattan, NY
2. 1937-41 Binghamton State Hospital, NY
3. 1941-46 Brooklyn, NY
4. 1946-48 Bronx, NY
5. 1948-50 Jackson Heights, Queens, NY
6. 1950-55 Flushing, Queens, NY
7. 1955 Fleming's farm, NY
8. 1955-59 Cornell U. Ithaca, NY
9. 1956 Ancora State Hospital, NJ
10. 1959 Fr. Bragg, NC
11. 1959-63 Indiana U Bloomington, IN

? 1963-65 Army (classified)

12. 1965-66 MGRU, Hammersmith Hospital, London, England
13. 1966-69 Univ. of Maryland, Catonsville, MD
14. 1969-present Univ. of Arizona, Tucson, AZ
15. 1976-77 Pasteur Institute, Paris, France
16. Summers Univ. of Lausanne, Lausanne, Switzerland
17. Summers Univ. of Cambridge, England
18. 1999-05 Our farm, Applegate, OR

Permissions to reproduce the following:

1. The figure on page 91 is reproduced from: Mendelson, N. H., and J. J. Thwaites. 1993. Bacterial macrofibers: multicellular chiral structures. ASM News. vol 59, No. 1, pp. 25-30. Reproduced with permission from the American Society for Microbiology.

2. The figure on page 94 is reproduced from: Mendelson, N. H. 1988. Regulation of *Bacillus subtilis* macrofiber twist development by D-cycloserine. J. Bacteriol. **170***:* 2336-2343. Permission granted by the Americal Society for Microbiology.

3. The figure on page 186 is reproduced from: Mendelson, N. H., J. E. Sarlls, and J. J. Thwaites. 2001. Motions caused by the growth of *Bacillus subtilis* macrofibres in fluid medium results in new forms of movement of the multicellular structures over solid surfaces. Microbiology **147**: 929-937. Premission granted by the Society for General Microbiology.

4. The comment, "and all the children are above average" that appears on page 158 is quoted from Garrison Keillor who wrote, "where all the women are strong, all the men are good looking, and all the children are above average." His writing is a registered trademark with the copyright office since 1997. The part reproduced here is with permission from Prairie Home Productions, LLC.

5. The discussion on page 210 dealing with a recent New Yorker cartoon pertains to the work of Charles Barsotti that appeared in The New Yorker on 3/7/2005. The publisher has informed me that permission to reproduce what has been said is not required in view of the fact that the cartoon itself is not reproduced.